Diversity of Life

Developed at
The Lawrence Hall of Science,
University of California, Berkeley
Published and distributed by
Delta Education,
a member of the School Specialty Family

© 2015 by The Regents of the University of California. All rights reserved. No part of this book may be reproduced or transmitted in any form or by any means, electronic or mechanical, including photocopying or recording, or by any information storage and retrieval system, without permission in writing from the publisher.

1465668
978-1-62571-173-1
Printing 2 — 4/2016
Quad/Graphics, Versailles, KY

Table of Contents

Readings

Investigation 1: What Is Life?
Characteristics of Life on Earth 3

Investigation 2: The Microscope
The History of the Microscope 8

Investigation 3: The Cell
The Amazing Paramecium 10
Cells 14

Investigation 4: Domains
Bacteria around Us 20
Bacteria: The Bad, the Good, and the
    New Frontiers 26

Investigation 5: Plants: The Vascular System
The Water-Conservation Problem 31
Water, Light, and Energy 35

Investigation 6: Plant Reproduction and Growth
Breeding Salt-Tolerant Wheat 40
Seeds on the Move 43
The Making of a New Plant: A Story about
    Sexual Reproduction 49

Investigation 7: Insects
Those Amazing Insects 51

Investigation 8: Diversity of Life
Biodiversity at Home and Abroad 59
Viruses: Living or Nonliving? 63

Images and Data 67

References
Science Safety Rules 98
Glossary 99
Index 104

Characteristics of Life on Earth

What Is Life?

It's not too difficult to tell that some things are alive. Dogs chasing tennis balls are alive. Birds chattering in a tree are alive. Fish swimming around the plants in an aquarium are alive. In fact, animals are the first things we learn to recognize as **living**.

Things that are alive, like the animals described above, are called **organisms**. Any living thing is an organism. But not all organisms are animals. In the photos to the right, the fruiting tree is alive, and the plants in the aquarium are alive.

It's not always easy to tell that plants are alive because they don't move around, breathe, eat, or make sounds. Even so, they are alive, and there are ways to figure out that they are living things.

Living, Dead, and Nonliving

One way to look at the question *What is life?* is to think about what makes life come to an end. Every living organism dies after a period of time. An organism is **dead** when it is no longer alive. A fish out of water will die after a short period of time. The fish is still there, it is still made out of the same materials, and it still looks the same as it did when it was living in the water, but it is no longer alive. And this is important—something can be dead only if it once lived. A rock can never be dead because a rock was never alive. We describe a rock as **nonliving**.

Living organisms can be described in terms of two sets of characteristics. One is the needs or requirements that all organisms have to satisfy to stay alive. The second is the **functions** that all organisms perform.

What Do Living Organisms Need?

What do you need to stay alive? It has been said that a person can live about 3 minutes without air, about 3 days without water, and about 3 weeks without **food**. People need air, water, and food to stay alive.

You breathe air to stay alive. When you breathe in, you bring oxygen into your lungs, where it dissolves into your blood. When you breathe out, carbon dioxide, carbon monoxide, and other **waste** gases leave your body and go into the air. The process of moving gases into and out of your body is called **gas exchange**. Birds do it, bees do it, lizards, fish, baboons, stink bugs, and even trees do it. All living organisms engage in gas exchange, and the most common gases exchanged are oxygen and carbon dioxide.

A stink bug

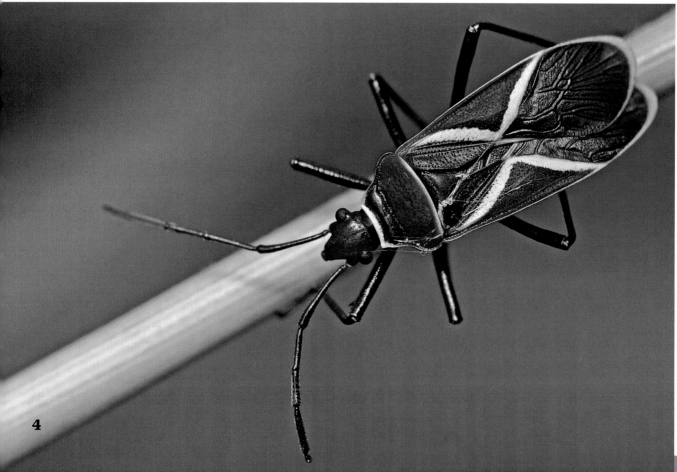

You drink water to stay alive. Even if you don't actually drink pure water, there is water in the **fruit**, vegetables, soft drinks, milk, and everything else you eat and drink. Water is essential for life as we know it on Earth. It's just that simple: all living organisms need water.

You eat food to stay alive. Food provides **energy**. Energy is required to make things happen. You can't move, breathe, see, hear, think, or do anything else without energy. All living organisms use energy to live.

The process of living creates by-products that are of no use to the organism. In fact, many by-products are dangerous to the organism if they are allowed to build up. For this reason, organisms must get rid of waste products. These might be gases, liquids, or solids. All living organisms eliminate waste.

It is a universal truth that everything has to be somewhere. That somewhere for an organism is its **environment**. Every organism lives in an environment that is suitable to fulfill its needs. Organisms have **adaptations** that allow them to live in their environment, or **habitat**.

The ocean and lakes are suitable environments for fish which have adaptations such as gills and fins. The desert is a suitable environment for scorpions, the forest for maple trees, fresh water and moist soil for **paramecia**, and so on.

Maple trees

A scorpion has adaptations to live in a hot and dry environment.

If the environment is not suitable, an organism will not survive. Some organisms form protective **spores** or capsules to survive unfavorable times. These spores do not appear to be living. They are **dormant**. But when suitable circumstances exist, they suddenly start to exhibit the characteristics of life. They were always living, but now you can tell.

Five basic needs are common to all living organisms. They are the need for *gas exchange*, the need for *water*, the need for *energy (food)*, the need to *eliminate waste*, and the need for a *suitable environment*.

What Do Living Organisms Do?

Once an organism's basic needs are met, it gets on with the process of life. When things happen in the environment, organisms respond. All organisms respond to the environment.

The ocean fish swims away when the sea lion comes by, the scorpion scurries under a rock when the Sun heats up the ground, and the maple tree's leaves turn red and fall off in the autumn. These are all **responses** to the environment.

When organisms start life, they are small. As time passes, they get bigger. An increase in size is called **growth**. The chemical building blocks for growth come from food, water, and from the environment in the form of minerals. All organisms grow.

Organisms don't live forever. To ensure that the **species** (a kind of organism) survives, living organisms make new organisms of their kind. They **reproduce**. That's not to say every individual organism will reproduce, but every population of organisms reproduces to keep the species going.

All organisms do three things. They *respond to the environment*, they *grow*, and they *reproduce*. Anything that does not have the ability to do all three of these things is not an organism.

There is actually one more characteristic common to all living organisms. That characteristic is not discussed in this article, but will be introduced in the near future. Can you think of what that characteristic might be? It's true of you, it's true of turtles and beetles, it's true of elm trees and mosses, and it's true of all the tiny living organisms too small to see with the naked eye.

Sometimes it is difficult to decide if something is alive. A car driving down the road exchanges gases, and a washing machine needs water. A burning candle uses energy, and a fire gives off waste. A smoke alarm responds to the environment, clouds grow, and the US Mint produces new dollar bills all the time.

One characteristic, or even three or four, does not qualify an object to join the ranks of the living. In order to qualify as a living organism an object must meet all eight criteria.

> **Think Questions**
>
> 1. What is an organism?
> 2. What are the *basic needs* of all living organisms?
> 3. What *functions* are performed by all living organisms?
> 4. Why do you think movement is not considered a characteristic of life?
> 5. Under what circumstances might a living organism not appear living?
> 6. What is the difference between living, nonliving, and dead?

A turtle and a beetle live in different environments, but they both respond to their own specific environment.

Investigation 1: What Is Life?

The History of the Microscope

It is not known when the first person picked up a piece of clear, curved material and found that it had **magnifying power** that made things look bigger. Roman books from the first century CE speak of "magnifying glasses," indicating that the Romans had some knowledge of what a *lens* does. By 1000 CE, people used glass spheres, called reading stones, that magnified text. The earliest simple **microscope** was just a tube with a lens at the top. It probably magnified no more than ten times, but was helpful for viewing tiny critters such as fleas, so it earned the nickname "flea glass." In 1590, when the first **compound microscope** was constructed, scientists were, for the first time, able to view the microscopic world. Improvements to existing designs and advances in technology continue to allow humans to see the smallest organisms in greater and greater detail (and even view substances at the atomic level!).

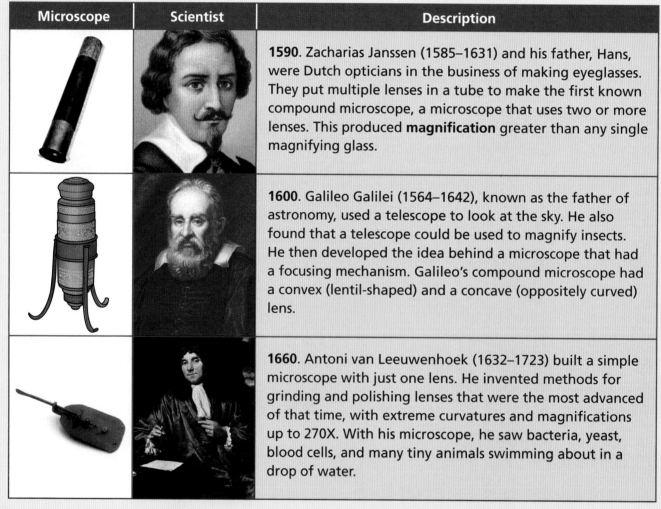

Microscope	Scientist	Description
		1590. Zacharias Janssen (1585–1631) and his father, Hans, were Dutch opticians in the business of making eyeglasses. They put multiple lenses in a tube to make the first known compound microscope, a microscope that uses two or more lenses. This produced **magnification** greater than any single magnifying glass.
		1600. Galileo Galilei (1564–1642), known as the father of astronomy, used a telescope to look at the sky. He also found that a telescope could be used to magnify insects. He then developed the idea behind a microscope that had a focusing mechanism. Galileo's compound microscope had a convex (lentil-shaped) and a concave (oppositely curved) lens.
		1660. Antoni van Leeuwenhoek (1632–1723) built a simple microscope with just one lens. He invented methods for grinding and polishing lenses that were the most advanced of that time, with extreme curvatures and magnifications up to 270X. With his microscope, he saw bacteria, yeast, blood cells, and many tiny animals swimming about in a drop of water.

Microscope	Scientist	Description
		1660. Robert Hooke (1635–1703) was an English contemporary of Leeuwenhoek who made a copy of Leeuwenhoek's light microscope and then improved upon the design. He looked at cork and noticed cells in it. Hooke confirmed Leeuwenhoek's discovery of the existence of tiny organisms in a drop of water. He wrote a book called *Micrographia* that officially documented many of the observations made through his microscope.
		1930. There were many microscope improvements after the 1600s, but the next big breakthrough came when Frits Zernike (1888–1966) invented the phase-contrast microscope. Until then, cell structures were made visible by staining, a process that killed the cells. The phase-contrast microscope made it possible to study living cells. Zernike won the Nobel Prize in Physics in 1953 for his work.
		1930. Ernst Ruska (1906–1988) and Max Knoll (1897–1969) coinvented the electron microscope. A visible-light microscope cannot be used for objects smaller than half the wavelength of visible light, about 0.275 micrometers (μm). To see smaller particles, we must use a different sort of "illumination," one with a wavelength shorter than visible light. An electron microscope speeds up electrons until their wavelength is only 100,000th that of visible light, making it possible to view objects as small as the diameter of an **atom**. Ruska won the Nobel Prize in Physics in 1986.
		1980. Gerd Binnig (1947–) and Heinrich Rohrer (1933–2013) invented the scanning tunneling microscope. It gives three-dimensional images of objects down to the atomic level. Binnig and Rohrer also won the Nobel Prize in Physics in 1986. The scanning tunneling microscope is one of the most powerful microscopes to date.

Investigation 2: The Microscope

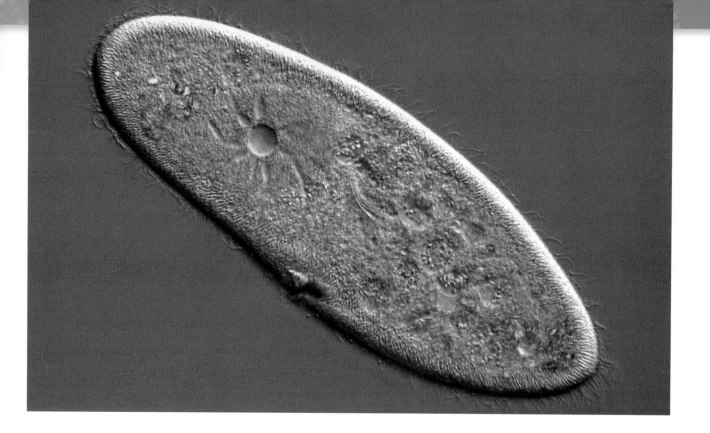

The Amazing Paramecium

Paramecia (pair•uh•ME•see•uh) are oval-shaped **single-celled organisms**, also called unicellular organisms. They are members of a large group of tiny organisms called **protists**. There are over 50,000 different kinds of protists. There are more different kinds of protists than all the different kinds of mammals, amphibians, fish, and birds combined.

We can be fairly certain that the first person to observe paramecia was Antoni van Leeuwenhoek (1632–1723), a Dutch biologist. In the mid-1600s, he spent a lot of time looking at things through his simple microscopes. Leeuwenhoek reported tiny objects swimming around in drops of water. He called them "animalcules," thinking they were tiny microscopic animals. We now know that protists are not, however, animals. They are single-celled organisms that live independently.

In plants, which are made of many **cells**, individual cells might specialize in making food or moving water. In animals, groups of cells might specialize in getting rid of waste, digesting food, or sensing the environment. In single-celled protists, a single cell must do all of the things that are done by the coordinated efforts of many cells in a plant

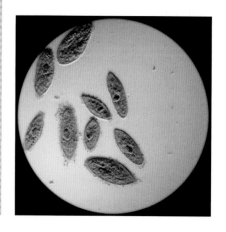

or animal. Each protist has the ability to respond to its environment, obtain food, exchange gases, get rid of waste, grow, reproduce, and use water.

When you use a microscope at 400X, you can see several kinds of **cell structures** called **organelles** inside the paramecium. Organelles are the paramecium's "guts." Cell structures have specific jobs that enable the paramecium to meet the needs of life and carry out life's functions. You have **organs**, such as a heart and kidneys, that do specific jobs in your body. The paramecium has **vacuoles** and **mitochondria**.

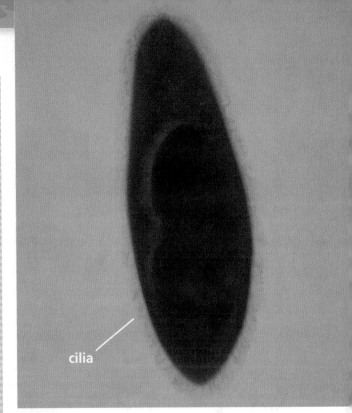

Special techniques make a paramecium's cilia visible.

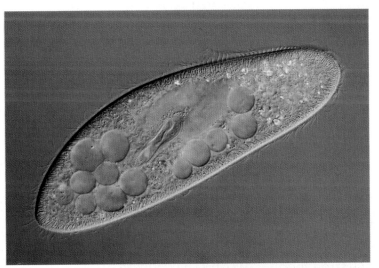

A paramecium viewed through a microscope

Paramecia are covered by rows of microscopic hairlike structures called **cilia**. *Cili* means small hair. Cilia move back and forth in a wavelike motion to move the paramecium through the water. Cilia are short, giving the paramecium a crew-cut look, and so fine that they are difficult to see even with a microscope at 400X. Cilia move water around the paramecium. If you watch closely, you might see tiny particles of debris moving in the water close to the paramecium. From this movement you can infer (figure out) that the cilia are moving, even if you cannot see them.

What Holds the Paramecium Together?

When you looked at a paramecium in class, you probably noticed shapes and textures inside the cell. There must be something like skin surrounding the cell, keeping the paramecium together. The paramecium's "skin" is called the **cell membrane**. Every cell has one, whether it is a free-living protist or a cell in a larger organism.

The membrane is one of the most important structures of the cell. The membrane defines the cell and keeps its structures and fluids on the inside, and everything else on the outside. If the cell membrane breaks, the cell quickly dies. A few materials, like water, oxygen, and carbon dioxide, can pass through the membrane, but most other materials cannot. So how does the paramecium get the food and other nutrients it needs to stay alive? How do nutrients get into the cell?

How Do They Eat?

Single-celled organisms don't have mouths that open to take in food the way animals do. Instead, paramecia have a fold in the membrane, called the **oral groove**, for taking in food. This fold runs most of the length of one side of the cell. When the cilia move back and forth, they swish materials in the water into the oral groove.

If the material is nutritious, the sides of the oral groove fold over the food and pinch it off in a closed packet called a food vacuole. The food vacuole moves around inside the paramecium. When paramecia consume red-dyed **yeast**, you can see the circular red food vacuoles right through the cell membrane.

The food is broken down by **digestive enzymes** while it is inside the food vacuole. The digested food moves out through the walls of the vacuole, so that it is available to the other parts of the cell for energy and to make new cell parts.

As the food is digested, the food vacuole becomes smaller and smaller. Finally, when the cell has digested all the valuable nutrients from the food, the vacuole moves to the cell membrane, and the leftover waste is dumped out of the cell through the membrane.

The paramecium takes food in and dumps waste out without ever opening the cell membrane to expose the inside of the cell to the outside environment.

Paramecia Drink, Too

Cells are constantly taking in water from their surroundings. All cells need a continuous supply of water in order to produce energy, repair worn parts, and do all the essential processes inside the cell. But with water constantly entering the cell, it would seem in danger of bursting before long. It has to rid itself of extra water.

The paramecium has **contractile vacuoles** in both ends of the cell. These organelles collect the extra water in the cell, along with some of the waste, and dump it out of the cell. They function very much like the kidneys in your body. You may have seen contractile vacuoles in the paramecia you studied. They look like little clear circles. They grow larger for several seconds and then suddenly become small again as the water is dumped out through a tiny pore in the cell membrane.

A contractile vacuole in this paramecium can be seen before and after eliminating waste.

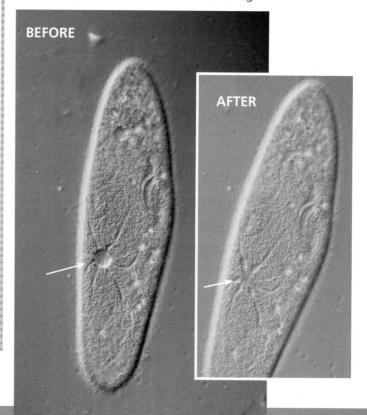

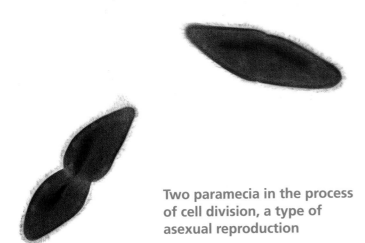

Two paramecia in the process of cell division, a type of asexual reproduction

Response to the Environment

Paramecia swim constantly, searching for food. One of the few times they stop is when they are feeding. They usually avoid cold or hot areas, or chemicals that would harm them, by swimming away from the danger area. Sometimes their **behavior** is funny to watch. You may have observed them swim straight on until they bump smack into something, then back up, turn, and swim off in another direction. Not altogether graceful, but effective.

Reproduction

Most of the time, paramecia reproduce by cell division, a form of **asexual reproduction**. They grow larger and make duplicates of all the organelles in the cell. When the cell reaches a certain size, it pinches together in the middle and splits in two, as shown in the image above. Each new cell is an exact copy of the original, but half as big. They are called **daughter cells**, and the original is called the mother cell. Even though they are called mother and daughter, they are not females. Unlike most plants and animals, there are no male and female protists, and **sexual reproduction** happens only rarely.

After the mother cell divides, it no longer exists. It has become two new paramecia! The new daughter cells immediately start doing the things all organisms do. They take in food and water and expel waste chemicals and gases. The food provides the energy for life and the building materials to grow. As the paramecia zip through their watery environment, they are constantly responding to food and hazardous conditions to improve their chances of survival.

The lives of paramecia and humans are as different as lives can be. But, as different as we are from these tiny protists, it is amazing to think about the many ways in which we are similar. The characteristics of life tie all organisms on Earth together.

Think Questions

1. Why is the cell membrane important?
2. What are two functions of the cilia?
3. What are the functions of the contractile vacuole?
4. What would happen to the paramecium if the contractile vacuoles stopped working?

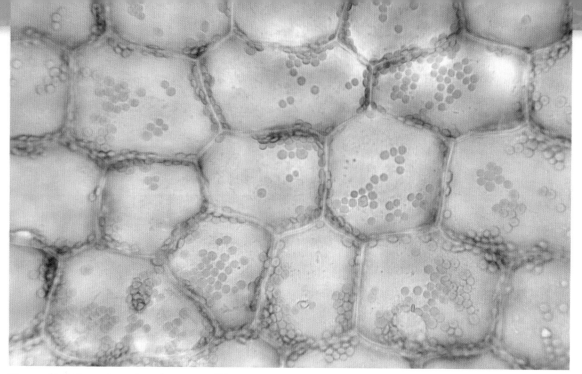

The cells of an *Anacharis* or Brazilian waterweed plant (*Egeria densa*)

Cells

"Why should I care about cells?" That's a reasonable question. You can't see them, and you might not have heard much about them before, so why do they matter?

Well, for one, all life on Earth exists as cells.

"Wait," you say, "*I am certainly not a cell.*" And you are 100 percent correct. But you are made of cells, such as nerve cells, liver cells, lung cells, blood cells, muscle cells, and many more.

Not only that, *every* living thing is made of cells. Some organisms, including **insects**, trees, worms, mushrooms, and the neighborhood cat, are **multicellular organisms**. That means they are made of millions, even trillions, of cells. In fact, the adult human body is made up of anywhere from 60 to 90 trillion cells. If you lined up all the cells in a human body end to end, you could circle Earth more than four times!

Most organisms, however, are single-celled. They consist of one cell. Organisms such as **bacteria**, **archaea**, paramecia, and other protists are all made of one cell. That's amazing! How could a single microscopic cell be a living thing? It begs the question: are cells living? What do you think?

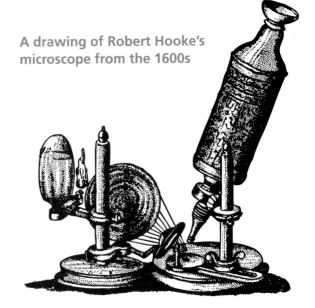

A drawing of Robert Hooke's microscope from the 1600s

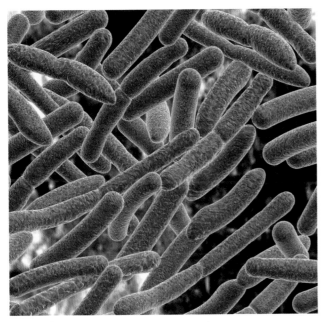

Bacteria

The invention of microscopes led to the discovery of cells. In 1665, Robert Hooke (1635–1703) looked at a sample of cork under the microscope. What he saw in the **field of view** looked to him like a bunch of tiny rooms. He drew the rooms to **scale** and called them cells, because they reminded him of rows of tiny rooms in a monastery, which were called cells.

All cells exhibit *all* the characteristics of life, even if they are part of a larger organism. Each cell carries out all the work necessary to sustain life. Before we figure out how they do this, let's step back for a moment to take a look at how cells were discovered.

The Discovery of Cells

Until the 1600s, no one had any idea that all life is made of cells. There was no way to see them. The technology didn't exist until eyeglass makers in Europe figured out that if they put several lenses together, they could magnify what they were looking at through the lenses. This discovery led to the invention of the first simple microscopes.

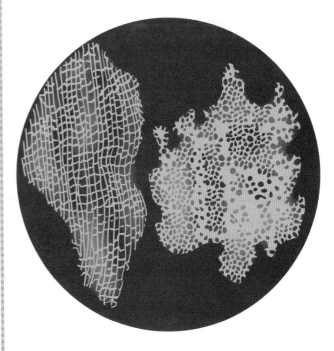

An illustration of the cork cells Hooke observed

Investigation 3: The Cell **15**

Around the same time, another scientist, Antoni van Leeuwenhoek (1632–1723), looked at samples of what appeared to be clear pond water. He was amazed to find that there were tiny things swimming around. His definition of life was not as sophisticated as the one we are currently using, but he concluded correctly that those swimming things were living organisms. He soon found them everywhere, even in his own mouth. He called the organisms animalcules, which means "little animals."

The Cell Theory

It wasn't long before scientific observations led to the conclusion that cells are the basic units of life. In the 1830s, biologists such as Matthias Schleiden (1804–1881) and Theodor Schwann (1810–1882) concluded that all plants and animals are made of cells. Soon it was confirmed that new cells come only from the division of existing cells. These conclusions led scientists to summarize their findings in what is called the cell theory. The cell theory states,

1. All living things are made up of one or more cells.
2. Cells are the basic units of structure and function in living things.
3. All living cells come from existing cells.

 Write the three statements of the cell theory in your science notebook.

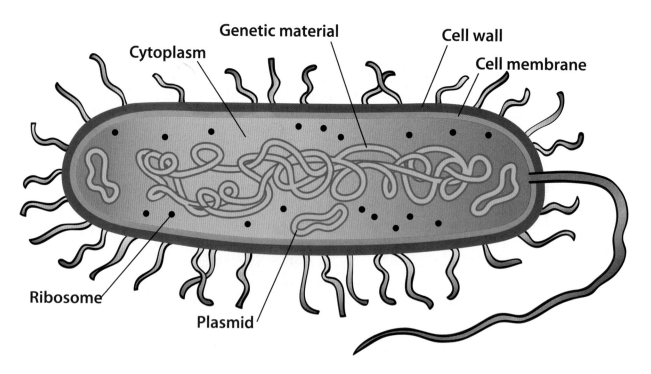

A bacterium cell

How Do Cells Carry Out Life's Functions?

Cells come in many shapes and sizes, some as small as 0.2 micrometers (μm) across, and some as large as 10 centimeters (cm) across. Some, as we have said, are individual living organisms, while some are the smallest living parts of much larger multicellular organisms.

All cells are separated from their environment by a cell membrane, a porous flexible structure that allows some things in and keeps other things out. This boundary keeps the inside fluid part of the cell, the **cytoplasm**, contained.

Cells such as bacteria and archaea are called **prokaryotes**. There is very little apparent organization of the materials inside the cell membrane. In fact, their **genetic material** in the form of deoxyribonucleic acid (DNA) or ribonucleic acid (RNA), simply floats in the cytoplasm. This is the main characteristic that defines prokaryotes. Prokaryotic cells carry out the business of life with very few specific cell structures.

Organisms with more complex cells (including all multicellular organisms) are called **eukaryotes**. In eukaryotic cells, the cytoplasm contains cell structures called organelles, meaning "little organs." Just as the human body is made up of organs and **organ systems**, which take care of life's functions, eukaryotic cells are made up of organelles, each of which has a job to do. Let's take a look at how cells carry out the functions necessary to sustain life.

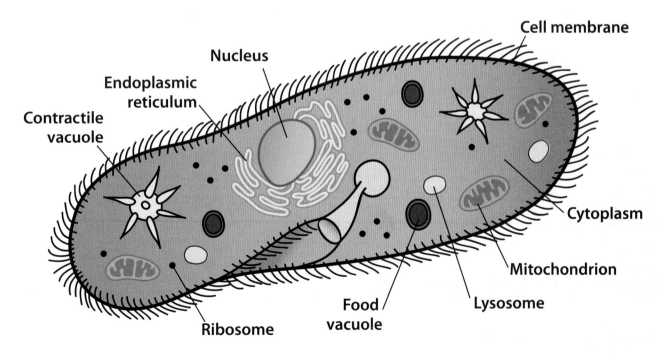

A protist cell

Investigation 3: The Cell 17

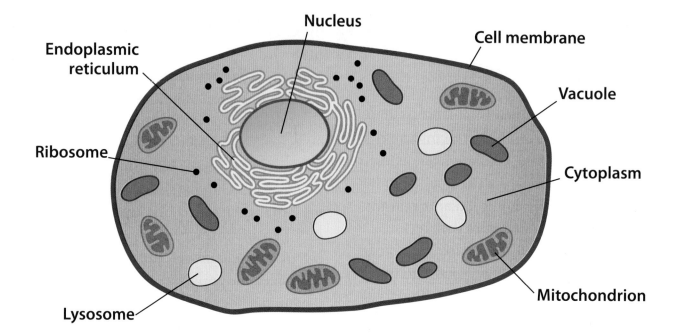

An animal cell

All cells exchange gases. Gases like carbon dioxide and oxygen move through the cell membrane.

All cells need water. In fact, all cells are found in an **aquatic** environment, a water-based fluid. Just like gases, water also flows back and forth across the cell membrane. Water is necessary for all chemical processes that happen in cells.

All cells need food. Plant cells make their own food in organelles called **chloroplasts**, using the Sun's energy, water, and carbon dioxide. Single-celled paramecia eat other **microorganisms**. Humans eat vegetables, fruits, and meat. Some bacteria consume sulfur. In eukaryotic cells, this food is transformed into usable energy by an organelle called the **mitochondrion**.

All cells eliminate waste. The **lysosome** is the animal cell's waste disposal system. Plant cells use chemicals for digestion in the central vacuole before discarding waste through the cell membrane.

All cells reproduce. The **nucleus** in eukaryotic cells contains the genetic information that drives cell division. And even though prokaryotes don't have a nucleus, they still have genetic material.

All cells grow. Several cell structures are involved in making proteins, which are used in building cell structures. They include **ribosomes**, which are sometimes free and, in eukaryotic cells, sometimes attached to a structure called the rough **endoplasmic reticulum**.

Look at the four cell images. Why do you think they have different cell structures?

18

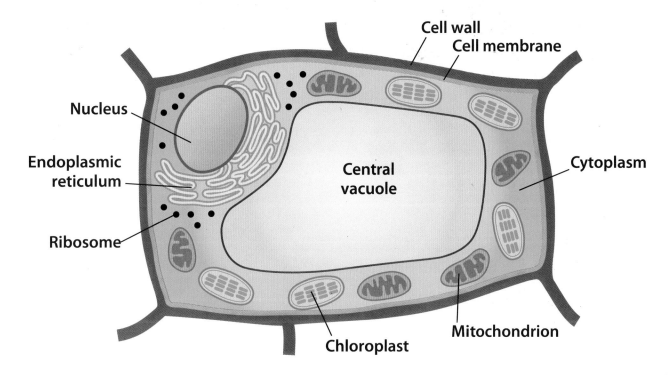

A plant cell

All cells respond to their environment. Paramecia swim around in search of food. They move away from cold or hot areas. Other kinds of cells respond to chemicals in their environment.

All cells need a suitable environment. Some cells form thick walls called cysts to protect them if an environment becomes stressful. If the amount of stress decreases, the cysts can break open, and the cells resume their lives. If the environment around other cells becomes toxic, they will die.

Why Should I Care?

Perhaps it's obvious by now. You, your friends, your family members, and every other living thing on Earth are composed of one or more cells. If cells did not exist, you would not exist!

Think Questions

1. How did changes in technology lead to the discovery of cells?
2. Describe how eukaryotic cells and prokaryotic cells are similar and how they are different.
3. Why are cells considered one of the characteristics of life?

Investigation 3: The Cell

Stromatolites, fossil rocks formed by the activity of ancient bacteria

Bacteria around Us

There are bacteria in my belly button, in my mouth, and on my toes? Are you serious? Yes! Bacteria are found everywhere—in and on our bodies and the bodies of every other organism, in the air we breathe, in polar ice, in water, in soil, in food, in the ocean, in dust, and even in fossil rocks. Bacteria are the most abundant organisms on Earth. It is impossible, even with the technology we have today, to know exactly how many bacteria exist, or even how many different species of bacteria exist. But it is estimated that there are five million trillion trillion (that's a 5 with 30 zeroes after it, or 5×10^{30}) bacteria on Earth. If we were to weigh them all, their mass would be greater than that of all the animals and plants on Earth combined.

Bacteria have lived on Earth for a very long time. Scientists have found fossils of bacteria in rocks in Western Australia (shown in the picture above) that are more than 3.5 billion years old. Some 2 billion years ago, bacteria started using energy from the Sun to make food. These bacteria produced oxygen as a waste product. The oxygen built up in Earth's atmosphere and became a key factor that led to the origin of multicellular organisms, including humans. Exploratory missions have been equipped by the National Aeronautics and Space Administration (NASA) to search for similar kinds of fossils in rocks on Mars as part of the quest to detect **evidence** of life there.

Bacteria Basics

Bacteria are very tiny single-celled organisms. The typical *Escherichia coli* bacterium (usually referred to as *E. coli*) is only about 2 micrometers (μm) (0.02 millimeters (mm)) across.

Because of their tiny size, bacteria were unknown to science until the late 1600s. While the effects of bacteria were observable, and masses of bacteria, called **colonies** or **cultures**, could be seen, the individual organisms themselves were a mystery until microscopes powerful enough to see them were developed. In 1676, Antoni van Leeuwenhoek was the first scientist to see bacteria through a microscope and to describe them. He found them in scrapings from human mouths. He used his own mouth, the mouths of two women (probably his wife and daughter), and the mouths of two old men who had never brushed their teeth.

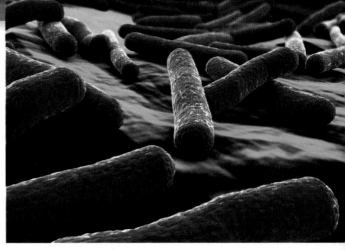

E. coli bacteria magnified

If Leeuwenhoek found bacteria in human mouths, where else do you think bacteria might be found?

Leeuwenhoek couldn't see inside the bacteria, but when scientists finally did, they found that bacteria looked quite different from other cells. Bacterial cells contain no organelles and no nucleus. Bacterial cells are called prokaryotes. (*Pro* means before, and *karyon* means kernel or nucleus.) But even though they look simple, they have their own way of organizing their **molecules**, and they are living organisms exhibiting the same characteristics of life as you and me.

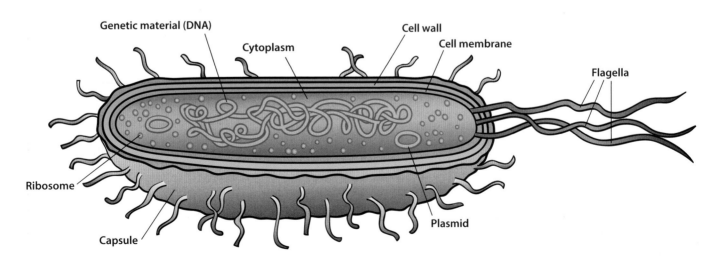

The makeup of a bacterial cell

Investigation 4: Domains

Bacteria are traditionally categorized based on their shape: rod, round, or spiral. They are also distinguished by how they appear when stained. More recently, many laboratories use DNA to identify bacteria.

Bacteria usually reproduce asexually. That means that they grow, make a copy of their DNA, and then split in two by making a **cell wall** and cell membrane between the two DNA molecules. This process is called binary fission (*binary* means two, and *fission* means split). If conditions are right, some bacteria, like *E. coli*, can divide every 20 minutes! Bacteria can also exchange genetic information in ways that resemble sexual reproduction. Two bacteria build a tube between them and exchange small bits of DNA called **plasmids**. Because each new cell gets new DNA, the bacteria cells acquire new traits. The ability to infect new hosts, resist an **antibiotic**, decompose or digest new materials, and many other traits can originate in this way.

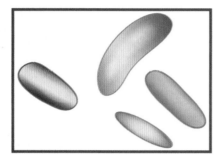
Rod-shaped bacteria

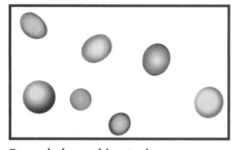
Round-shaped bacteria

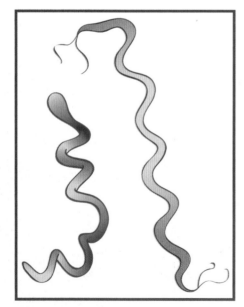
Spiral-shaped bacteria

 What other organism reproduces asexually? What are the advantages of asexual reproduction?

So Many! Everywhere!

Bacteria are found in every environment. Some species survive in boiling hot springs in Yellowstone National Park, while others live in environments that are very acidic, very alkaline, or high in sulfur. One of the Yellowstone bacteria, *Thermus aquaticus*, led to the discovery of a chemical that is useful in DNA fingerprinting, and the patent holder received a Nobel Prize and a million dollars!

Bacteria are found in polar and glacial ice samples that have been undisturbed and buried for thousands of years. When snow falls, bacteria can attach to the snowflakes. As the snow accumulates and turns to ice, the bacteria get buried within the ice. Researchers have retrieved ice cores all over the world, and in each case, they find bacteria embedded in the ice. The bacteria are in a

A hot spring in Yellowstone National Park. The brilliant color is caused by the bacteria *Thermus aquaticus*.

dormant state, but researchers have been able to "revive" some of them, including some that have been frozen in the ice for more than 100,000 years. NASA is interested in this research because of the possibility that bacteria might survive frozen in places like the moons of Jupiter and Saturn.

Bacteria have also been found high in the atmosphere above Earth, in the stratosphere, where temperatures average lower than −56 degrees Celsius (°C) and radiation levels are high. A bit lower in the atmosphere, bacteria can affect cloud formation and the chemistry of the air. Bacteria can be blown into the air from soil and water. One study found more than 1,800 different kinds of bacteria in air sampled over San Antonio, Texas. In another study, scientists found that bacteria can travel by wind over long distances. They found more than 1,000 different kinds of bacteria that traveled in the atmosphere from China, over the Pacific Ocean, to the United States.

The bacteria blown into the atmosphere also can fall to Earth when it rains. Researchers are finding that bacteria in the atmosphere can act as particles around which snow, hail, and raindrops form. The bacteria

Even icebergs harbor dormant bacteria.

Investigation 4: Domains

can fall to Earth with the precipitation. Bacteria are also part of the natural **ecosystems** of lakes and streams, helping break down complex molecules into nutrients for other organisms.

Bacteria exist in great numbers in soil. It is estimated that there might be 100 million bacteria in 1 gram (g) of soil, and that 92 to 94 percent of all bacteria live underground. Bacteria have been discovered in solid rock 3 kilometers (km) below the surface of Earth. These bacteria feed on chemicals in the rocks or chemicals released from rock as it breaks down. While *E. coli* can divide every 20 minutes, it may take these underground bacteria a thousand years to reproduce.

Not only are bacteria found high in the atmosphere, and in lakes, streams, and soil, they are also found deep in the ocean. Scientists are discovering that many diverse communities of bacteria live at various depths and parts of the ocean. In a 10-year project, scientists from 80 countries collaborated to try to inventory the many kinds of bacteria living in the world's ocean. Their estimate is that there might be more than a *billion* different kinds of bacteria living in the ocean.

Bacteria by the Numbers
- Scientists have estimated that there are 100 times more bacteria in the ocean than there are stars in the known universe.
- Six liters (L) of seawater contain more bacteria than there are people on Earth. That means a mouthful of seawater might contain a million bacteria!
- If you were to stack the same number of pennies as there are bacteria on Earth, the stack would stretch a trillion light-years high.

A Zoo of My Own

Bacteria live in and on our bodies. So let's go back to our belly buttons. Are my belly-button bacteria like my friend's belly button bacteria? As it turns out, the answer is probably not. According to the Belly Button Biodiversity Project, your belly-button bacteria are unique. You have twice as many kinds of bacteria in your belly button as there are species of ants or birds in North America. That's more than 2,300 different kinds! (What do you think they eat?)

Other scientists have found that the bacteria on our fingertips are also fairly unique. Researchers were able to get cultures from keyboards and determine who used those keyboards (a new tool for crime scene investigators?). So you, or at least your bacteria, are unique. And while your skin is home to trillions of individual bacteria, your mouth and gut have the greatest diversity of bacteria.

At the beginning of the article you wrote other places bacteria might be found. Add to your list!

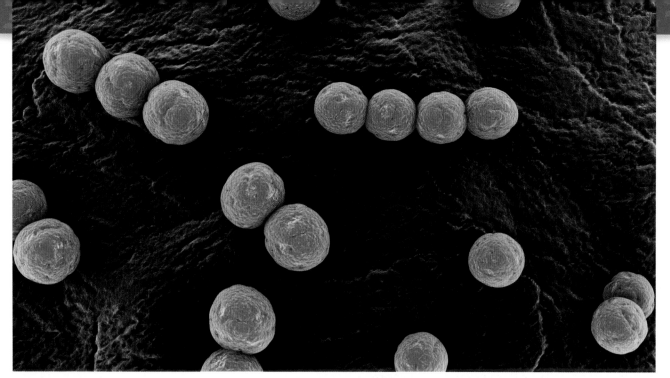

Streptococcus bacteria

Scientists still don't know where all the bacteria live in our bodies and what role they play in keeping us healthy or making us sick. Some bacteria might help people maintain a healthy weight, and others might determine how a human responds to antibiotics.

Projects like the Human Microbiome Project are attempting to answer those questions about bacteria. The study of our interaction with the thousands of different kinds of microbes in and on us is a fascinating new area of human biology and medicine. Scientists hope to discover a lot of information about how we can stay healthy. Where might researchers plan the next safari? Armpits!

Are We Really Human?

Here are some numbers that might make you rethink what it means to be human.

- Each of us is home to about 100 trillion microscopic life forms (that includes bacteria, yeasts, viruses, and protists). The number of bacterial cells in the human body is roughly equal to the number of human cells!

- In a 100-kilogram (kg) adult, all these microorganisms would weigh 1 to 3 kg. That's more than your brain weighs!

**So, are we really human?
Or are we just a habitat for bacteria and other microorganisms?**

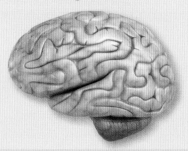

Think Questions

1. What are some examples of suitable environments for bacteria?

2. How might bacteria avoid dying if an environment is not suitable?

3. How are bacterial cells different from plant and animal cells?

4. Describe how bacteria reproduce. How can bacteria acquire new characteristics?

Investigation 4: Domains

Bacteria: The Bad, the Good, and the New Frontiers

The Bad

An innocent cow walks around a grassy field. What could possibly be wrong with this picture? What do cows do besides walk, eat, drink, and moo? Hold onto that question.

Escherichia coli, or *E. coli* as it is generally known, is one of the most common bacteria on Earth. These bacteria are found in the intestines of all warm-blooded animals, including humans; and in soil, rivers, lakes, and beach sand. Most strains of *E. coli* are benign (harmless).

One particular strain of *E. coli*, *E. coli* O157:H7, lives quite comfortably in the intestines of cows, and the cows don't mind this at all. It's not too hard to figure out how *E. coli* O157:H7 get into lakes and soil, and sometimes into hamburger meat! Unfortunately, this strain of *E. coli* that is harmless to cows is potentially deadly for humans. We use water-quality tests to look for bacteria like *E. coli* to make sure that our water isn't contaminated, and we process meat so that the cow's intestinal bacteria don't end up on the meat sold in the grocery store.

Many different kinds of bacteria normally live in the human body and cause no problem. But when the number of one kind of bacteria gets large, or the balance of the different kinds of bacteria changes, there can be problems. Bacteria were responsible for about 40 percent of **food-borne illnesses** between the years 2006 and 2010. In addition to *E. coli,* other bacteria have also been responsible for food-poisoning episodes. They include *Listeria* in hot dogs, deli meat, and cantaloupes; *Salmonella* in eggs, peppers, turkey, and products containing peanuts; and *Staphylococcus* in meat.

Peanut butter

Every year, about 3,000 people in the United States die and another 128,000 people end up in the hospital because of food-borne illnesses. This costs our economy about 7 billion dollars. That is why it is so important to cook meat thoroughly, drink juice that has been pasteurized, and wash fresh fruits and vegetables before eating them.

Bacteria are harmful to humans when they damage or kill human cells in large numbers or in some way prevent the cells from functioning. Bacteria can damage cells by releasing toxins either directly into nearby cells or into the bloodstream to reach cells and **tissues** in other parts of the body. Bacteria can use dead cells as a food source. This is called a bacterial infection. Bacteria that are most harmful to humans can reproduce quickly and overwhelm the immune system.

Bacteria have caused disease in humans throughout history. They have killed off entire populations and caused more deaths than wars. What are some of the worst bacterial outbreaks in history?

- The black death, or plague, is caused by *Yersinia pestis.* It is carried by the fleas on small animals like rats, or transmitted from one person to another. Plague is estimated to have killed more than half the human population of Europe in the 1300s.

- Tuberculosis (also called TB) is caused by *Mycobacterium tuberculosis.* It spreads from one person to another through the air and attacks the lungs. In 17th-century Europe, TB killed about one in every seven people. Today it still kills more than 2 million people per year worldwide. It is becoming resistant to existing treatments, meaning that outbreaks might become worse in the future.

- *Rickettsia prowazekii* causes typhus and is transmitted by lice. Outbreaks of typhus sometimes changed the outcome of wars and thus history, and have been responsible for the death of millions.

- *Vibrio cholerae* is carried by contaminated food and water. It causes cholera, which leads to intense diarrhea and vomiting. One of the worst outbreaks in US history happened in the 1800s. It started in India, and eventually appeared in New York, New York and New Orleans, Louisiana. From there it spread westward as settlers traveled the Oregon Trail.

Investigation 4: Domains

While now rare in the United States, cholera still kills more than 100,000 people each year around the world, especially in developing countries and areas struck by natural disasters, such as Haiti after the 2010 earthquake.

One disease you might have heard of in the news recently is Lyme disease. It is caused by *Borrelia burgdorferi* bacteria, and it spreads to humans by deer ticks. A bull's-eye rash is characteristic of the disease. It is one of the fastest-growing infectious diseases in the United States, though as of 2012 it is mostly found in the Northeast and upper Midwest. While the bacterium that causes Lyme disease has been around for a long time, climate change, cutting down forests, and building houses in areas that have a lot of deer ticks have all contributed to the increase of the disease.

Bacteria may seem scary. But humans have some help in the form of medicines that can kill many types of harmful bacteria. These medicines are called antibiotics (*anti* = against, *bio* = life) and have been effective in eliminating or controlling many disease-causing bacteria.

What bacterial outbreaks or food recalls have you heard of in the news lately?

Antibiotic Resistance

One of the big problems facing us today is that some harmful bacteria are becoming resistant to these traditional antibiotics. Resistant means that antibiotics can't kill off these bacteria completely. One example is *Staphylococcus aureus*, the bacteria that cause most hospital infections. Methicillin is a strong antibiotic used to treat these infections. In recent years, this bacterium has become increasingly resistant to methicillin. Another example is multi-drug resistant tuberculosis.

How do bacteria become resistant to antibiotics? Some individuals in a population of bacteria are naturally resistant, and when a population of bacteria is exposed to an antibiotic, those that are resistant survive to reproduce. All of the offspring are resistant. Additionally, mutations and gene-sharing (via plasmids) can help bacteria survive when exposed to antibiotics. Mutations

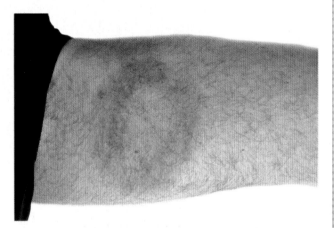
A bull's-eye rash

A deer tick

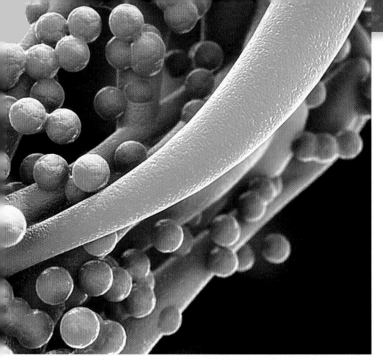

Staphylococcus aureus is usually harmless unless it enters deeper parts of the body.

(random changes in genetic material) occur naturally as bacteria reproduce. Bacteria can also gain new genes when they share genetic material with other bacteria, even bacteria from a different species. When we overuse antibiotics or use them incorrectly, it is possible that we unintentionally increase the survival rate of resistant bacteria.

The Good

We've been focusing on the "bad," but in reality, without bacteria, Earth would probably not be a very nice place to live. Bacteria are **decomposers**, which means that they break down dead organisms and waste into water, gases, and minerals. Can you imagine an Earth buried in dead things? Without bacteria, that's what it would be like! Bacteria change nitrogen in the air into a form that plants can use to grow and flourish and are essential to maintaining the ecological balance of the planet.

Bacteria make chemicals that humans use for other purposes. Botulinum is a deadly toxin produced by the bacteria *Clostridium botulinum*. But one form of the toxin can be injected into muscles in the face, smoothing out wrinkles. The toxin is also used to treat people with severe muscle cramps, saving them from terrible pain. Scientists are developing ways to modify bacteria to make them produce chemicals that are useful medicines. *E. coli* is being genetically modified to produce chemicals like insulin and human growth hormone, which are useful in medicine. And oil-eating bacteria can help break down environmentally harmful oil spills.

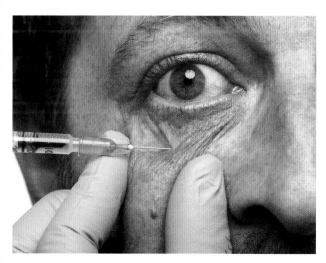

Botulinum injections

Speaking of breaking things down, the community of bacteria and other organisms living in human intestines is necessary for digestion of food. The bacteria in our digestive tract make some of the vitamins we need and help us ward off other bacteria that might make us sick. Scientists think that because we humans have changed the variety of food we eat, we've changed the diversity of bacteria living inside of us. These changes in the bacterial community could contribute to weight gain and obesity.

Bacteria also help make some of the foods we eat. They are used to make salami, pepperoni, sour cream, chocolate, cider, soy sauce, pickles, coffee, vinegar, yogurt, cheese, hot sauce, kimchi, poi, and sourdough bread, to name a few examples. Bacteria are also used to make thickening agents for sauces and salad dressings, and in gluten-free baking.

Bacteria are used in the processing of coffee beans.

The holes in Swiss cheese are caused by bacterial activity.

The New Frontiers

What lies ahead for those trying to learn more about bacteria, the most abundant organisms on Earth? Currently, the most common species of bacteria on Earth is thought to be *E. coli*. It is perhaps the most studied organism on Earth. *E. coli* grows easily in the lab and reproduces rapidly, about every 20 minutes. This means that scientists can document its **evolution** (change in species over time) without waiting for millions of years. *E. coli* has helped us understand how proteins in our own body are made, how cells age, how DNA is organized, and how genes work.

The science of microbiology plays a role in almost every field of science, and much of our microbial world is yet to be discovered and understood. As our ability to find and distinguish different bacteria increases, so too will the number of known species. And as the number of known species increases, perhaps we'll find other bacteria more common than *E. coli*. Or perhaps there are bacteria to be discovered on other planets, moons, or other objects in the solar system. There's plenty of opportunity for a career in this fascinating field!

Think Questions

1. How can you help prevent food poisoning?

2. Why is it important not to overuse antibiotics?

3. How are bacteria helpful in the human body?

4. *E. coli* is both dangerous and helpful to humans. What are some ways this bacterium is dangerous and some ways it is helpful?

The Water-Conservation Problem

Plants face a curious problem. They need carbon dioxide and water to make food and to sustain life in their cells. Where do they get these substances? Carbon dioxide comes into a leaf through the **stomata** (singular stoma). This usually happens during the day, so that carbon dioxide is available for **photosynthesis** while the Sun shines. But water exits a leaf through the stomata. So if a plant keeps the stomata open on a bright, sunny day to let carbon dioxide in, it loses water. If a plant loses too much water, the cells shrink and the plant wilts. If the wilted plant isn't watered, chemical reactions in the cells stop and the cells die. How do plants balance their need for carbon dioxide with their need to conserve water? Plants have remarkable ways to address this problem.

Stomata Control

The most important way leaves manage water loss is via the stomata. Each stoma is surrounded by two **guard cells**. Fully hydrated (water-filled) guard cells are banana shaped. A pair of guard cells holds each stoma open to allow gases like water vapor and

Investigation 5: Plants: The Vascular System

carbon dioxide to enter and exit. When the cells in the leaf start to dehydrate, the guard cells also lose water and flatten out. The result is that the stomata close and water loss is reduced significantly.

What is interesting is that the size of the stomata does not affect the entrance of carbon dioxide as much as it does the exit of water. If a stoma is partially closed, the exit of water is slowed down, but carbon dioxide can still enter the leaf, and photosynthesis continues. However, when the stomata are totally closed during particularly hot or arid conditions, even carbon dioxide can no longer enter the leaf. Whatever carbon dioxide remains inside the leaf is quickly used up, and photosynthesis stops. Sometimes for the plant it is more important in the short run to close the stomata to conserve water in the cells than it is to make **sugar** through photosynthesis.

Leaf Adaptations

Leaves have specific adaptations that also help them conserve water. Most leaves have more stomata on the bottom surface than on the top surface. This protects the leaf because less water evaporates from open stomata on the shadier, cooler underside of the leaf.

The **cuticle**, a layer of waxy material on the surface of the leaf, reduces the amount of water that evaporates out of the cells and into the air. In dry climates, the cuticle on the leaves of some plants is very thick. Additionally, some plants have very thick leaves that can hold a lot of water. Because the leaves are so thick, most of the water in the leaf is farther from the

The spines on a saguaro are modified leaves.

surface and the stomata, helping conserve water. Many desert plants have very small leaves, resulting in a smaller surface area available for **transpiration**. In fact, the spines on cacti are actually modified leaves! However, the small surface also cuts down on photosynthesis. This usually works out fine, because sunlight is rarely a problem for desert plants. And plants like the paloverde tree and cacti have chloroplasts in their stems. Photosynthesis can occur in those green stems as well as in the leaves.

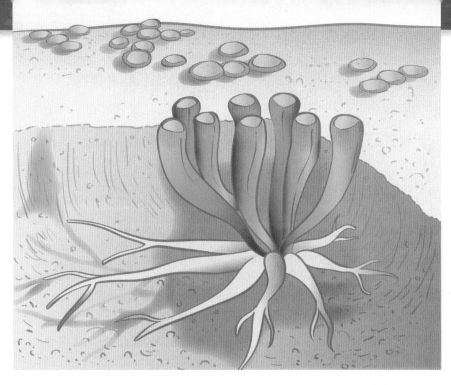

Fenestraria rhopalophylla

One desert plant has a remarkable way of conserving water. *Fenestraria rhopalophylla* of South Africa grows underground with only the tips of the leaves above ground. The leaves are very thick, and the center part is packed with water-filled cells. The cells surrounding these water-tank cells contain chloroplasts. The leaves have a clear "window" at the exposed tip. Sunshine passes through the window into the center of the leaf to the underground cells containing the chloroplasts, where photosynthesis occurs. The cells are protected from being dried out, and they also get plenty of sunlight.

Leaves That Collect Water

Some leaves actually collect the water the plant needs. For example, the redwoods along the California and Oregon coast obtain about half the water they need from the fog that comes off the ocean. The tiny droplets of fog collect on the short, thin needles of the redwood trees and drip off. During one night of heavy fog, as much water can drip off a redwood as during a drenching rain. This

Redwood trees

The stomata in the pineapple plant's leaves open only at night to reduce the loss of water.

keeps the trees and the plants under them alive during the summer months when there is little rain, but plenty of fog. The redwoods' needles also soak up some of that fog, rather than waiting for it to drip off, go into the soil, and be drawn up by the **roots**. A shortcut indeed. A redwood fulfills about 10 percent of its water needs in this fashion.

Other plants growing in dry climates have fuzzy leaves that collect moisture from dew. The fuzz increases the surface area of the leaf, creating more area for vapor to condense into liquid water. Dew collected by hairy-leaf plants keeps the soil moist.

Another Kind of Photosynthesis

Here's another great water-saving trick. Some plants don't open their stomata during the day at all. They open them only at night, letting carbon dioxide in, but losing much less water. But there's a problem. Because it's night, there is no sunlight, so the plant can't photosynthesize. Instead, the plant converts that carbon dioxide into other molecules, effectively trapping the carbon. When the Sun comes up, the plant tightly shuts its stomata so no more water can be lost. The trapped carbon inside the leaves is converted back to carbon dioxide, and photosynthesis starts right up.

Where would you expect to find plants that use this kind of photosynthesis?

Summary

Plants have **evolved** many amazing adaptations to balance their need for carbon dioxide with their need to reduce transpiration and conserve water. Plants have many strategies, such as different kinds of leaves, the location of the stomata, and using a different kind of photosynthesis. Plants' strategies are as varied as they are!

Think Questions

1. Why do plants need both carbon dioxide and water?
2. What are the basic needs of all living organisms?
3. Water transpires through the leaves of a plant. What are several strategies that plants use to decrease the amount of transpiration?

Water, Light, and Energy

It was your chore to water the houseplants, but you forgot! When you finally remembered, it was too late. The plants looked brown and shriveled. Plants need water in order to live. "Needs water" is on the list of life requirements for *all* life.

Have you ever seen houseplants wilt, lose their healthy green color, and die if they don't get enough light? "Needs light" isn't on our list. Is light one of the requirements for all life? Should we add it to the list or can we figure out how plants' need for light fits into the list we already have?

 Record in your notebook your ideas about whether "needs light" should be added to the list of requirements for life.

All Plant Cells Need Water

The answer to our question about light has a somewhat unexpected starting point: water. Plants use water for many things: to transport minerals to all their cells, to dissolve substances to make them available for chemical reactions, to cool off in the heat of the day, to give them shape, and to grow. Plant cells are filled with cytoplasm, which is mostly water.

Investigation 5: Plants: The Vascular System

Plants get the water they need from the soil. **Root hairs** take up water and pass it into hollow tubes, which make up **xylem** (ZY-lem) tissue, part of the plant's **vascular system**.

Xylem tubes in celery

Veins in celery leaves

As we saw in the celery investigation, the xylem carries water up through the stem to the smaller **veins** in the leaves. In this way, all the cells in a plant get water and the minerals from the soil that come along with it. Xylem tubes are made of the cell walls of dead xylem cells, which are connected end to end in the stems, like long straws. The tubes form an extensive system of pipes that can end up being extremely long and complex, especially if we are talking about a tree as tall and massive as this giant sequoia.

A giant sequoia tree

Transpiration

What happens when water finally reaches the leaves? What did you observe when you put a bag around a leafy twig? Water! Where did that water come from? It came from the plant. It left the plant through the leaves as water vapor and entered the atmosphere. This process is called transpiration. Water vapor (lots of it!) exits the leaves through small pores called stomata (*stoma* = mouth). Guard cells open and close the stomata to control the movement of gases, including water vapor, into and out of the leaf.

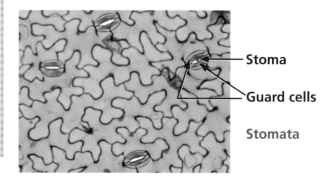

Stomata

Photosynthesis

You have probably heard that plants and **algae** (and some bacteria) make their own food. How do they do it, and what do water and light have to do with it? The process of photosynthesis is the answer. Thinking about photosynthesis will help us answer our original question about light. Water is involved in this food-making process in at least two ways.

Remember that water dissolves substances to make them available for chemical reactions. One of those substances is carbon dioxide (CO_2) gas. The stomata open during the day to allow gas exchange, allowing carbon dioxide from the atmosphere to enter.

The carbon dioxide dissolves in water in the spaces surrounding the cells. The carbon dioxide enters the nearby cells, where it becomes one of the building blocks of sugar, which provides food energy for the plant. And this could provide food energy for any other living thing that eats the plant. So water makes carbon dioxide available.

The second reason that water is important is that water itself is the *other* building block of sugar. Water combines chemically with carbon dioxide to make sugar.

Only one thing is missing from this equation. What do you think it is? The other thing plants need to survive is light.

The overall simplified chemical reaction can be expressed like this.

$$6CO_2 + 6H_2O + \text{light energy} \longrightarrow C_6H_{12}O_6 + 6O_2$$

Carbon dioxide + water + light energy *makes* **sugar + oxygen**

There is something very interesting about this reaction. It occurs only in the chloroplasts, the green organelles that you first encountered in **elodea** leaf cells. You can't just dissolve some carbon dioxide, throw it together with water, hit it with light energy, and expect to make sugars. The reaction only happens in the chloroplasts where there is a green chemical pigment called **chlorophyll** that allows the plant to capture and convert light energy into the chemical bonds in sugar.

In order to make their own food, plants need water, carbon dioxide, light energy in the form of sunlight, and chlorophyll. The process is called photosynthesis, which makes sense because *photo* means light, and *synthesis* means putting together.

Light and water are both critical in the process of photosynthesis.

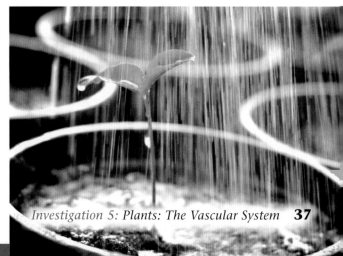

Investigation 5: Plants: The Vascular System

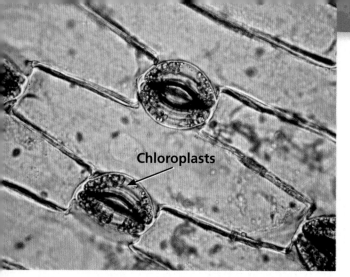

Chloroplasts

Chloroplasts are even found in guard cells.

Aerobic Cellular Respiration

Now we know how plant cells that have chloroplasts make food. But look at the plant cells below. They are from the root of an elderberry plant and they do not have chloroplasts. How do they get the food they need? The food that was made in the cells with chlorophyll must get to the root cells somehow. Right alongside the xylem is another part of the plant's vascular system called **phloem** (FLO–em) tubes. These tubes carry sugar from the leaves to all the other cells of the plant.

All the cells of a plant get food delivered via the phloem. But the energy stored in sugar is not in a form plants can use to grow, repair damaged tissue, or make new **structures**. In order to change sugar to a form that cells *can* use, plants need oxygen.

Cross section of an elderberry root. Note the lack of chloroplasts in the cells surrounding the xylem and phloem.

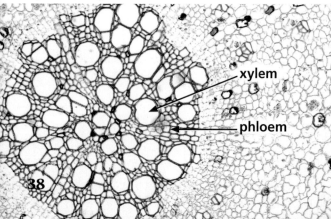

xylem

phloem

You know that in photosynthesis, plants use carbon dioxide and give off oxygen. But plants need oxygen, too. Like most living cells, plant cells use oxygen to transform sugars into a usable form of energy. Oxygen and sugar, in the form of a molecule called glucose, combine to release energy, and carbon dioxide and water are given off as waste by-products. This reaction happens in every plant cell, every animal cell, and almost all other single-celled or multicellular organisms' cells.

The chemical reaction can be expressed like this.

$$C_6H_{12}O_6 + 6O_2 \rightarrow 6CO_2 + 6H_2O + energy$$

Glucose + oxygen *makes* carbon dioxide + water + energy.

In eukaryotic cells, this process happens in the mitochondria and is called **aerobic cellular respiration** (*aerobic* means it uses oxygen). Notice anything interesting? Compare the equation for photosynthesis and the equation for cellular respiration. The sugar molecules are on opposite sides of the equations, and the light energy was turned into usable energy for the plant. Almost all organisms rely on aerobic cellular respiration to convert glucose into usable energy. But only photosynthetic organisms can capture the Sun's energy to create sugars. How do other organisms get sugars? All other organisms must eat photosynthetic organisms, such as plants, or eat organisms that did so.

 How are photosynthesis and aerobic cellular respiration alike? How are they different?

Summary

- Plants need water. They pull water up from soil using transpiration. Plants transport water to all cells, using tubes (xylem) that are part of their vascular system. Water vapor exits leaves through stomata, which are opened and closed by guard cells.
- Plants need food. They make their own food out of water and carbon dioxide, using light and chlorophyll in a process called photosynthesis. The food they make is a sugar (glucose), which stores energy.
- Plants transport sugar to all cells, using tubes (phloem) that are part of the vascular system.
- Plants (and almost all life forms) use aerobic cellular respiration to change food (sugar in the form of glucose) into usable energy to perform all of life's functions.

We don't need to add "needs light" to our list of life requirements. The light plants need is a part of "needs food." So be sure to water those plants and make sure that they get the light they need!

Think Questions

1. Explain why water is necessary for plants to make food.
2. How do all the cells in a plant get the water they need? Explain.
3. Do all plant and animal cells photosynthesize? Explain.
4. Do all plant and animal cells use aerobic cellular respiration? Explain.
5. Is light a requirement for life? Explain.

A wheat field

Breeding Salt-Tolerant Wheat

Australia has a big problem. The land is facing a farming crisis. Wheat is Australia's most important food crop, but the farmland where wheat is grown is becoming more and more affected by salt (sodium chloride, in particular).

Soil Salinity

Australia is not the only place in the world where agriculture is harmed by high concentrations of salt in the soil. Salt is found naturally in many soils around the world. But when high levels of soil **salinity** (salt concentration in soil) are found in agricultural land, the salt can affect how the land produces crops. About 20 percent of the world's farming soils and half of the world's irrigated lands are affected by high soil salinity. These high levels can be caused by drought or deforestation. Farmers irrigate soil to grow more crops to feed people yet naturally-occurring salts in the water can build up in the soil. Today, soil salinity is one of the most important **environmental factors** affecting the success or failure of food crops.

How does soil salinity affect plants? Salt can prevent a plant from taking in the water it needs for photosynthesis and thus for growth. Lack of water can stunt a plant's growth and decrease production of the **seeds** and other parts we depend upon for food.

If the soil salinity is too high, water may actually leave the plant's cells, causing the plant to die from dehydration, even if there is moisture in the soil.

Wheat dying from high soil salinity

Salt Tolerance in Grains

Can some plants grow in saline soils? Remember what you learned about how corn, barley, oats, and wheat grow in salty conditions. How did these seeds **germinate** and grow in salty environments compared with the seeds in the fresh water **control** dishes? If seeds do germinate in salty environments, they are considered **salt tolerant**.

Plants that are salt tolerant have **genetic factors** (genes in their DNA) that help the plants in different ways. One gene in modern bread wheat (*Triticum aestivum*) prevents salt from entering the roots or traveling up the xylem, thus protecting the plant's leaves. A gene in barley (*Hordeum vulgare*) causes the plant to capture and store salt in its vacuoles, so that it doesn't reach the leaves.

Unfortunately, these crops are not the main food crops for much of the world's population. Durum wheat (*Triticum turgidum*) is the main grain in many parts of the world. It is used to make pasta, flat breads, bulgur, and couscous. It grows well in semiarid climates like North Africa, and in parts of India, Europe, and the Middle East. But durum wheat is vulnerable to high soil salinity. Scientists in Australia isolated a "salt-tolerant gene" in einkorn wheat (*Triticum monococcum*), an ancient relative of durum wheat. They found that this gene makes a protein that keeps sodium from entering the shoots of the wheat. The scientists were able to breed a new variety of durum wheat that incorporated the salt-tolerant gene from the einkorn wheat. The new plant grows successfully in salty fields.

Couscous

Pasta

Durum wheat

Flat bread

Investigation 6: Plant Reproduction and Growth

The human population is predicted to reach 9 billion people by 2050, with the demand for food expected to be twice what it is today. As the occurrence of drought increases and sea levels rise, soil salinity is increasing in many vital farming areas. This means that the land available to grow wheat is decreasing. By engineering wheat crops to be more salt tolerant, these creative scientists have made it possible to grow wheat in those salty soils. They have made great strides toward ensuring future food security not only for the people of Australia, but for billions of people worldwide.

> **Think Questions**
>
> 1. How does soil salinity (an environmental factor) affect plants?
> 2. How do genetic factors allow some plants to be more salt tolerant?
> 3. How are scientists making durum wheat more salt tolerant? Why is this important?

Earth's population is expected to reach 9 billion people by 2050.

Seeds on the Move

Plants are everywhere around us. Often the types of plants that you see in an area help define a location, such as the majestic redwoods of California, the brilliant foliage of the hardwoods in New England, and the giant saguaro cacti of Arizona. Where did these plants come from?

Once a plant puts down roots, it is anchored for good. It can't move closer to a water source or seek a place with more direct access to sunlight. In the earliest phase of its life, however, a plant can move. Most plants grow from seeds, and because seeds are small and self-contained, they easily move from one place to another. During the seed phase of their lives, plants expand their range and colonize new territory.

But there is one problem with this plan. Seeds don't have legs, fins, or wings. They can't move by themselves. If they are going to establish themselves in a new place, they need an agent to move them.

The process of spreading out from a starting place is called **dispersal**. Young plants often benefit from being some distance from the parent plant because they don't have to compete with the larger, well-established plant for resources. The methods used by plants to disperse their seeds are called **seed-dispersal strategies**, and the structures on the seeds that allow them to move are **seed-dispersal mechanisms**.

One strategy for seed dispersal is to produce a lot of seeds. Chances are, if a plant produces 10,000 seeds, a few of them will end up some distance from the parent. For instance, the Asian poppy produces huge numbers of small, smooth, round seeds. Most of them fall out

Investigation 6: Plant Reproduction and Growth

of the pod and end up quite close to the parent. Now and then, however, one might fall onto something sticky, like a little drop of sap. If a person, dog, or rodent happens to step on the seed, it might stick to a foot for a while and be carried a considerable distance before it falls off. If the new location is suitable for poppies, the plant has succeeded in expanding its range. A one in 10,000 chance of survival is not very promising, but in the long run it works, allowing the plant population to be successful.

Wind

Some plants use wind to disperse seeds. The seeds are usually very light and frequently have some kind of wind-catching mechanism, such as a sail, tuft, puffball, or parachute. Wind-borne seeds travel until the wind stops, or they snag on an obstacle, or they get soaked by rain or dew. Dandelion and milkweed plants produce tufted seeds that can travel for many kilometers before landing. Silver maple seeds come in pairs and look like wings. When a gust of wind shakes them loose from the tree, they fly along on wind currents and fall, spinning like a helicopter.

Maple seeds fall like helicopters.

The tumbleweed plant of the southwest United States uses a variation on the wind strategy. After producing seeds, the tumbleweed dies and breaks off from its roots. The dead plant is a light, nearly spherical, compact mass of branches and twigs covered with thousands of seeds. When the wind really starts to blow, the tumbleweed goes bounding and tumbling across the desert or prairie, leaving a trail of seeds in its wake. Seeds that fall in favorable areas can grow and develop into next year's tumbleweed crop. This strategy is so successful that the tumbleweed is incredibly invasive in environments where it is not native. The tumbleweed originated in Russia and was introduced to the American West in the 1870s. It is now considered a pest, as it can grow to the size of a small car and decreases the quality of agricultural lands.

A milkweed plant's tufted seeds

A tumbleweed contains thousands of seeds.

The edible part of a coconut is its seed.

Water

Plants that grow in or near the water often use floating as their seed-dispersal strategy. Floaters are usually pretty light, with a shell covering the fruit that is less dense than water. A waxy, watertight coating often covers the fruit inside.

The coconut palm is the champion when it comes to long-distance dispersal using water. Coconut palms are adapted to grow right on beaches. The trees may even grow out over the water and drop the fruit directly into the tide. More often the fruits drop on the beach, where they may be washed into the sea later by high tides or storms.

What you see in the grocery store is the coconut seed. It is huge. In fact, it's one of the largest in the world. The fruit of the coconut is even larger and is made of a very low-density fibrous material. A coconut can float on ocean currents for weeks before salt water penetrates the seed and ruins it. If it happens to wash up on a beach before it goes bad, it may germinate.

Many of the plants found on tropical islands were transported this way, and a stroll along the beach will yield a wide variety of floating seeds.

Some seeds are dispersed by water.

Coconut palms are adapted to grow right on beaches.

Investigation 6: Plant Reproduction and Growth **45**

A housecat covered with seeds

Animals

Animals participate in seed dispersal in many ways. Some plants use the piggyback strategy. Did you ever pet a housecat or dog and discover a burr or foxtail or other seed matted in its fur? Hooks, barbs, coils, and sticky stuff can make a seed cling to the fur, feathers, or feet of an animal that chances by. Once attached to an animal, the seed might travel a few meters or even kilometers before it falls off or is scratched free by the carrier. These hitchhikers are usually fairly small and light, and may be sticky like glue (saguaro cactus) or covered with a variety of hooks and spikes (bur clover, cockleburs, foxtails, and bull thistles).

Want to find out what kind of plants in your neighborhood use this strategy for dispersing seeds? Take an old pair of worn-out wool socks, pull them on over your shoes, and take a short walk through a dry field or a walkway that has high dead grass. Check out the socks after a while. Try to remove the seeds, and you will see how effective some of the dispersal mechanisms are for holding onto a host animal. You might go one step further and plant the old socks under a couple of centimeters of soil, water them, and see what comes up!

Another way seeds are dispersed is by animals eating the fruit that contains them. Some seeds pass completely through an animal's digestive tract unharmed. Such seeds have very durable seed coats, as does the black cherry. A magpie might swallow a cherry, fly to the next county while digesting the fruit from around the pit, and rid itself of the seed in a dropping several kilometers away. Birds and fruit bats have carried seeds between many of the small islands in the tropics in exactly this way.

Birds are great transporters of seeds.

A third way that animals aid in seed dispersal is by gathering and storing seeds for food. Squirrels are famous for burying acorns, peanuts, and other nuts in many places in preparation for winter. They are equally famous for forgetting where they buried them. These lost or forgotten seeds may sprout and grow when spring arrives. Ants also gather seeds for food and store them underground for later use. These seeds may also grow if not eaten.

A squirrel often forgets where it has stored seeds.

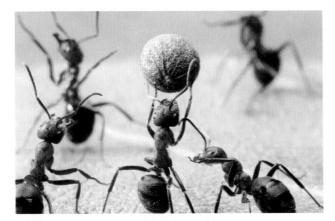

Ants store seeds underground.

Ejection

Some plants use the "heave-ho" method for dispersing their seeds. As bean pods dry on the parent plant, the pods twist and become brittle. When completely dry, they suddenly burst, and the bean seeds are thrown away from the parent. The wisteria plant is a champion in this technique, propelling seeds 20 meters (m) or more with a loud crack as the pods release their stored energy. Mistletoe is a parasitic plant that attaches to a tree limb and draws water from the host. When the seedpods mature, they burst and eject a soft, sticky seed up to 15 m away. If the seed hits another tree, it will stick and grow into a new mistletoe plant.

Bean pods burst when they dry out, releasing the seeds inside.

Investigation 6: Plant Reproduction and Growth

Combination

Some plants disperse seeds in more than one way. Cosmos flowers are an example. Each cosmos flower produces many seeds. Each seed has hooks that make it easy to stick on the fur of passing animals or the clothing of people and be dispersed widely. Many other seeds fall to the ground near the parent plant and may get buried in the soil. If conditions remain favorable there, the buried seeds will sprout where the parent plant grew the season before. If conditions have changed for the worse, seeds dispersed elsewhere may grow better in a new location that is more favorable for growth.

Back to the opening question . . . where did the plants come from? They came from all over. Some flew in, some were launched in, others rode in on the backs of animals, some were dropped as the leftovers from someone's supper, and a few might have floated in. Each plant is growing where it is because its parent had some form of seed-dispersal mechanism that worked.

With the successful dispersal of the seed, the new plant thrives, and the life cycle continues.

Think Questions

1. Why is seed dispersal important for a plant?

2. Do all seeds that have been dispersed from a plant come out of dormancy? Why or why not?

3. What are some of the seed-dispersal strategies you found in your seed search?

The Making of a New Plant: A Story about Sexual Reproduction

Flowering plants reproduce sexually. That is, new plants are formed with genetic information from two plants. What are the steps in the process?

Seeds contain an **embryo**, which is the living baby plant that has the potential to grow into a new plant. The **ovary** of the **flower** is where these seeds develop. Seeds start as **ovules**, tiny pre-seeds, in the ovary. The ovary is located at the base of the **pistil** in the center of the flower. Inside each ovule is a single, female sex cell, an **egg**.

A male **sperm** cell inside a **pollen** grain on top of the pistil must fuse (join) with a female egg cell deep inside the ovary at the base of another flower. How does it do this? The pollen is transported from one flower

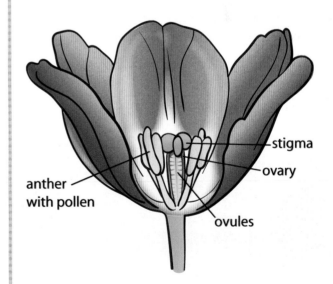

to another with the help of an animal, wind, or water. The delivery of the pollen is **pollination**. Shortly after landing on the

Investigation 6: Plant Reproduction and Growth

sticky **stigma**, the pollen grain performs an astonishing feat. It grows a long tube, like a root, down the length of the pistil and into the ovule. The sperm, which contains the male's genetic information, travels through the **pollen tube** into the ovule to fertilize the egg, which contains the female's genetic information.

After successful **fertilization**, the single cell divides, and each of those cells divides, and so on and so on, until the cells develop into an embryo like the one you saw in the seed during the seed dissection. Then development stops.

The parent plant supplies the resting embryo with a package of energy-rich food, the future **cotyledon**, and wraps the whole system in a weatherproof coat. The plant has produced a seed, the living package that will produce the next generation.

Some plants have flowers that produce a single seed, like a peach flower or a cherry blossom. In this case, the ovary contains only one ovule. Other plants, like green beans or apple trees, have flowers with maybe five to fifteen ovules in the ovary, and others, like tomato and watermelon flowers, have hundreds of ovules in the ovary. Each ovule has the potential to produce a new plant if it is fertilized. At the same time the fertilized ovule is developing into a seed, the ovary that holds the seed is developing into a fruit. The fruit is any structure that grows around the seeds and ensures the survival and success of the next generation. Familiar examples of fruits include grapes, lemons, cantaloupes, and pears. Scientifically speaking, a number of objects that we often refer to as vegetables are in fact fruits, including tomatoes, squash, beans, cucumbers, olives, peanuts, and eggplants. The general rule is that, if it has seeds, it is a fruit.

The fruit may fall from the plant, get eaten, or be carried away from the parent plant. If the seed inside the fruit finds a suitable location, it will germinate and grow into a new plant.

The watermelon fruit grows from the ovary of the female watermelon flower.

Those Amazing Insects

There are millions of different kinds of insects tucked into every imaginable niche on this planet. How can Earth provide a place for so many different kinds? Each insect has some structure or behavior that makes it different from other insects, and provides it with a unique way to get the resources it needs, find the space it requires, and reproduce its own kind. Some of the structures and behaviors that insects have evolved are really amazing, and there are undoubtedly many more that entomologists (scientists who study insects) have yet to discover.

A Madagascar hissing cockroach

What's That Sound?

An interesting characteristic that you are surely familiar with by now is the sound made by the Madagascar hissing cockroach. Scientists wondered why these roaches hiss

Investigation 7: Insects **51**

and what advantages they would gain from hissing. Did you notice what causes them to hiss? Often they let out a hiss when they are harassed by another animal. You may have heard a hiss when you picked one up. But do they hiss at other times?

Scientists noticed that roaches sometimes hiss when there is no threat from another animal. They also observed that males, and only males, hiss in the presence of a female. This led the scientists to think that hissing might be involved in mating behavior, perhaps used by males to establish their territory or to scare off other males.

Hissing cockroaches produce their hiss by forcing air out of the **spiracles** on both sides of their fourth segment. In order to test their ideas about hissing and mating behavior, scientists covered up the spiracles on the fourth segment of one male roach and left them open on another male.

Both male roaches were placed in a cage to see which would become the dominant male. The hissing male almost always became the boss.

In other experiments, scientists also found that the male that hissed the loudest almost always drove off the other males. When a hissing and a nonhissing male were in a cage with a female roach that was ready to mate, the male roach that could hiss fought off the roach that couldn't. The hissing male was more likely to mate with the female, and pass on his "hissing" genes. In fact, the females would not mate with a male that couldn't hiss.

Little Drummer Wasp

Humans have been cultivating and storing grain for thousands of years. Insects have been sharing the annual harvest of grain for thousands of years as well. One insect in particular, a type of weevil, chews a tiny hole in a kernel of wheat and lays a single egg inside. When the egg hatches, the weevil **larva** eats its way into the kernel of wheat and consumes the inside of the kernel until it is hollow. By this time the larva is ready to pupate, and some weeks later the next generation of weevil emerges. A tidy little lifestyle.

A wheat weevil

Rice weevils

It's never quite that simple, however. Also living in the same area is a tiny wasp whose larva eats the larva of the weevil. The female wasp lays one egg on the outside of a wheat kernel. When the egg hatches, the larva burrows into the kernel and devours the weevil living inside. But how does this wasp know which wheat kernels contain weevil larvae? They are sealed inside of the kernel, and there are literally millions of kernels to choose from.

The mother wasp crawls around on the outside of the kernels and uses her antennae like drumsticks to beat on the kernels! Just as an empty plastic bin sounds different than a full plastic bin when you bang on the outside, a hollow wheat kernel with a weevil inside sounds different to the wasp than a wheat kernel that has not been hollowed out. This curious and effective behavior allows the female wasp to leave her eggs where they have the highest probability of survival.

So the human behavior of harvesting and storing wheat increases the survival potential of both the weevil and the wasp. Do you see why?

Aphids on a branch

Reproduction Strategies

Aphids are tiny insects that are sometimes considered pests because they literally suck the life out of plants. Aphids have mouthparts shaped like a straw that they insert into a plant's phloem to tap the sugar-rich sap. If a large number of aphids descend on a plant, they can weaken or even kill it.

Aphids can reproduce very quickly because they give birth to live, fully developed babies. This is almost like being born an adult, except that the newborns are tiny.

Aphids can kill plants.

Newborn aphids can begin feeding and reproducing almost immediately.

When an aphid is born, it can begin feeding *and reproducing* almost immediately. Talk about a head start!

Young aphids are clones of their mother, which means that all aphids are female. There are no male aphids. Not only do aphids have asexual reproduction, they have also evolved another strategy for shortening the time between generations. Many aphids are born pregnant. Entomologists have dissected aphids under a microscope and found aphids ready to give birth to baby aphids that also had babies inside of them! It is easy to see how aphids could take over an entire field of plants in just days.

The Big Aphid Roundup

Imagine a life where you never drank water. Instead, every time you were thirsty you reached for a big liter-sized bottle of your favorite soft drink and gulped it down. Then you followed it with another and another and another . . . That's basically the life of an aphid.

Instead of drinking soft drinks all day, aphids drink the sugar-rich sap from plants. The aphids can't digest it all, so the extra sugar comes out the back of the aphids as a sticky sweet substance called honeydew.

At some point, this honeydew came to the attention of a particular kind of ant. In time, the ants came to rely on it as their only source of food. They developed an amazing way to guarantee a steady supply of honeydew for themselves. They created ranches! The aphids are their stock, and plants with a good supply of sap are their ranges.

Ants and aphids on a branch

54

Ladybugs feed on aphids.

Every morning, the ants round up the aphids and drive or carry them out to feed on plants. While they are feeding, the ants make sure that the aphids don't wander off or get rustled by an outsider. If a ladybird beetle (a voracious predator that eats six times its own weight in aphids every day, also known as a ladybug) comes by, the ants herd the aphids into small groups and defend their stock from attack. At the end of the day, the ants take the aphids back to the anthill, where they spend the night, and the whole thing happens again the next day.

In exchange for this protection and care from the ants, the aphids let the ants harvest the honeydew for food. This benefits the ants (because they get a reliable source of great food) and the aphids (because they get a place to live and protection from predators). If you ever get a chance to watch this happen, you'll be amazed at how much it resembles cowboys herding cattle on a ranch!

Aphids depend on protection from ants.

Ants working together

Pheromones: Insect Calling Cards

Ants are social insects. All social insects live together and rely on each other for survival. They work together to raise the young, construct and maintain the living quarters, defend the colony, and obtain food.

Acquiring food is always a challenge. To locate food, the foragers leave the colony and strike out into the environment, looking for something good to eat. They wander out without a plan and without a guidance system. The path they follow is random as they search here and there. If an ant runs across a scrap of seed or a grain of sugar, she will eat it, but her quest is for something good that is much more than she can eat herself. She is foraging for the whole colony.

When she does come across a dead moth, piece of fruit, or chunk of cheese, she pries off a crumb and heads back to the nest.

She keeps an antenna to the ground to pick up the faint scent laid down by other ants to guide her home. As she advances toward home, she lowers her **abdomen** to leave a tiny drop of a powerful chemical on the ground. This is a **pheromone**, a chemical message that other ants will follow to relocate the food source.

Ants work together to bring a piece of fruit to the colony.

56

Back at the nest, the ant shows off the sample of the prize she has located. The other foragers note the particular pheromone smell of the ant that brought in the food, and follow her smell back to the bounty. The other foragers retrace the path of the first one, each leaving additional pheromone markers on the path. In a short time, a wide stream of thousands of ants is hurrying in both directions over the invisible but smelly (if you are an ant) trail. Pheromones are very effective agents of communication for ants, both to assist with the business of the colony and to identify intruders from rival colonies.

Trail marking and identification are two ways ants use pheromones. Moths use pheromones in another way. Because moths are active at night, it's not so easy to find things. And one thing that must be found to ensure the survival of the species is a mate.

An anthill

Moths are nocturnal, or active at night. (*Saturnia pryi*)

The female moth uses a pheromone to attract a mate. The male responds. Many moths find a mate in the following way. When the time of year is right, a female moth flies to a tree limb or rock and starts to advertise her availability by releasing a bit of her irresistible pheromone perfume. Any male moth of her species up to 2 kilometers (km) downwind who happens to encounter a molecule or two of the heavenly scent on his antenna will start to fly toward its source. If he is fortunate enough to find the source, he may mate, reinforcing the effectiveness of the scent to bring the male and female moths together by passing the scent gene on to offspring.

Every insect has a story to tell about how it survives and reproduces. The stories above are just a few of the astonishing adaptations of insects. There are thousands more. For instance, how do mosquitoes find you in the dark when you are the only person around for miles? Why do moths fly around your porch light? What are the leaf-cutter ants doing underground with so many circles of leaf? Imagine how many more tales there are and how many more natural history stories there will be when scientists study all the species of insects.

Think Questions

1. When does a hissing cockroach hiss? How does the hissing benefit the cockroach?

2. The wheat weevil and the drummer wasp may benefit from the activities of humans. Can you think of another insect that may benefit from living around humans? Explain the benefit.

3. If you bother a wasp, you might find yourself pursued by the whole colony. How do you think they communicate to know who to chase?

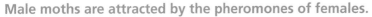

Male moths are attracted by the pheromones of females.

Biodiversity at Home and Abroad

Bioblitz: Biodiversity Close to Home

It is early morning. The Sun has not yet warmed the ground, but the excitement is clearly brewing as people begin to arrive. As participants sign in at the registration tent, it appears that some kind of race is about to begin. But you notice that the surrounding booths are filled with skulls, skins, and live animals and plants, and the tables are covered with laptops, field guides, data sheets, microscopes, and other equipment. What is going on? A bioblitz!

A student counts the organisms she found.

Investigation 8: Diversity of Life **59**

During a bioblitz, teams of scientists, amateur naturalists, and community members of all ages work together in a designated site to tally as many organisms as they can in a race against time: 24 hours. What's the point? A bioblitz brings attention to the incredible variety of life in urban areas such as city parks or schoolyards, and wilder spaces, such as national parks. As the list of organisms in an area grows, scientists can identify species that might not otherwise have been known, or decide if certain species need to be observed more closely over time.

E. O. Wilson, author and renowned ecologist

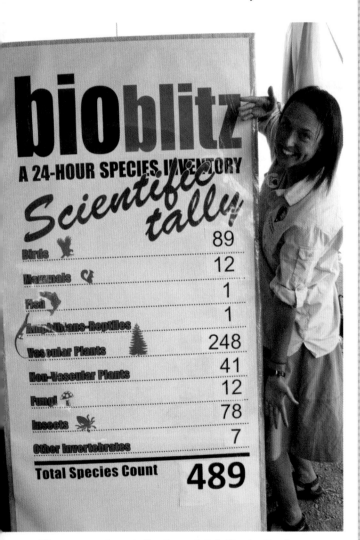

The organism tally for a bioblitz in Rocky Mountain National Park in Colorado

It is important to remember that these events cannot replace more technical monitoring. A one-day event cannot come close to documenting all the species present, and even repeating a bioblitz on the same day each year won't always track change over time, like more specific monitoring projects can. But a bioblitz is very useful for raising public awareness, involving the local community in understanding local **biodiversity** (*bio* = life; *diversity* = variety), connecting the public to science, and giving people a chance to learn naturalist techniques that anyone can use anywhere and anytime. At the same time, a bioblitz contributes unique data to the scientific field. As E. O. Wilson (1929–), a Harvard biology professor and renowned naturalist, explains, a bioblitz is "a wonderful treasure hunt, a scientific research program, and a generally wonderful outing among people accomplishing something."

Student scientists discover larvae under a rock.

Measuring Biodiversity

Measuring biodiversity is a complicated task. To measure the biodiversity in an area, you basically measure two things. The first thing is the number of different species of plants, animals, protists, bacteria, archaea, and **fungi** that exist in the area. The second thing is the number of organisms present. Scientists are discovering how the many species in an ecosystem (a system of organisms and environmental factors) rely on each other. Scientists have found that an ecosystem could have many different species present, but if a key player, such as a **pollinator**, goes missing, the entire system could be in trouble. Thus, biodiversity can be seen as a measure of the health of an ecosystem.

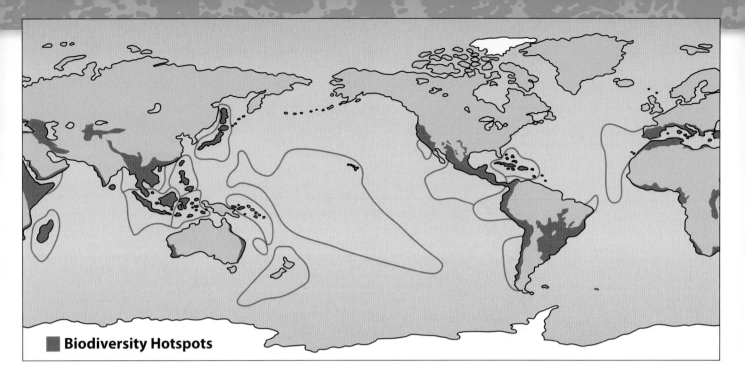

■ **Biodiversity Hotspots**

Biodiversity Hotspots

Although it is important to preserve biodiversity everywhere, areas called biodiversity hotspots harbor a high diversity of species and are under considerable threat by human impact. Accounting for 1.4 percent of the land surface on Earth, these hotspots are home to nearly 60 percent of the world's plant, bird, mammal, reptile, and amphibian species.

Many current biodiversity hotspots are in countries where large numbers of people live in poverty. Therefore it is essential to consider ways to maintain biodiversity while also improving the quality of life for local people. Efforts such as using fewer chemical pesticides, growing coffee in the shade of native trees, and having farmland with many types of crops can help accomplish these two goals. Ecotourism can also bring money to local people while supporting biodiversity preservation.

Biodiversity: The Future

The combination of climate change and

human development will continue to pose major challenges for the world's biodiversity in the coming century, but we humans can be creative and find ways to live more harmoniously with other species. Current and future scientists will help us do this important work. Perhaps participating in a bioblitz might be the ticket to your positive impact on this world. "Go as far as you can, [young scientists]," writes E. O. Wilson in his book, *Letters to a Young Scientist*. "The world needs you badly."

Go to FOSSweb to explore the resources on biodiversity and bioblitzes.

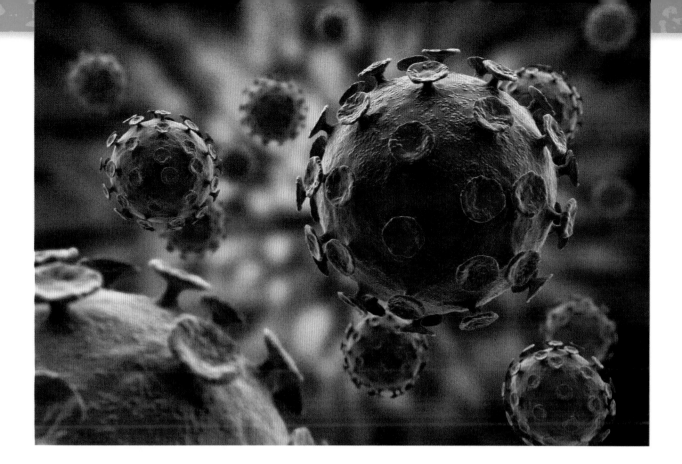

Viruses: Living or Nonliving?

At the beginning of the course, we posed the question, "How do you know if something is living?" Since that time you have learned a lot. Let's apply that knowledge to think about **viruses**. If you had to vote based on the evidence you currently have, would you say viruses are living or nonliving?

Perhaps we should do just a bit more investigating. As you think about how viruses have interacted with more familiar forms of life throughout life's history on Earth, consider how the new information here affects your conclusion.

Viruses have probably been around for billions of years. Billions? Yes, billions. But their origins are still debated by scientists.

Did they appear with and evolve with bacteria and other cells? Did they appear before? Did they evolve from single-celled organisms? The debate continues, but it helps to remember that humans discovered viruses only a little over 100 years ago. A virus was first observed only after the electron microscope was invented in the 1930s, providing the first technology that could let humans study viruses up close. There is a lot of work still to do to understand viruses. One thing we do know, however, is that there are a lot of viruses, more than 10^{31} (10 billion trillion trillion). That is more than all known life-forms combined, including bacteria, archaea, protists, fungi, plants, and animals!

Portrait of a Virus

Viruses come in a multitude of shapes and sizes. They are not made of cells. Instead they are made up of genetic material in the form of a nucleic acid (DNA or RNA) which is surrounded by a protective coat called an envelope. The genetic material is surrounded by a protective coat called an envelope.

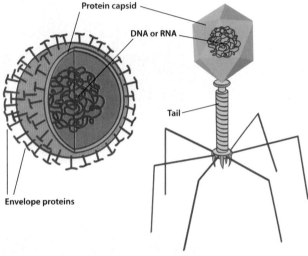

The structure of a virus

Because it is not a cell itself, in order to reproduce, the virus has to enter the cell of an organism, called a host cell. In some viruses, the envelope surrounds the whole virus and has protein molecules that stick out to help it attach to a host cell, so it can inject its genetic material. Other viruses have a protein tail that attaches to the host.

Part Virus?

Scientists have discovered that, amazingly, humans are part virus. The genetic material of viruses makes up 8 percent of human genetic material. What do all of these embedded genes do? Most of the viral genes have changed over time, so that they no longer have any influence, but one of them is responsible for making a protein that causes a fetus's placenta to stick to its mother's uterus. Without this essential viral gene, you might never have been born!

It appears that viral genetic material became part of the human genome over thousands and thousands of years of infections. That is, as viruses have infected humans, they sometimes inserted their DNA into our sex cells (eggs or sperm) and took over those cells' genetic machinery. Some of these human cells survived to pass on the viral DNA to the next human generation. In some cases, such as the one mentioned above, these genetic changes have proved to be helpful.

Viral Diseases

Influenza (the flu), smallpox, measles, AIDS, rabies, Ebola, Marburg, West Nile, Hanta, yellow fever, herpes, hepatitis, warts, polio, mumps, rubella, chicken pox, shingles, SARS, cold sores, and norovirus are just a few of the human diseases and infections caused by viruses. Some, like smallpox, have led to millions of deaths. Others, like chicken pox, stay in your body for years, emerging later in life to cause a painful rash called shingles. Some, like the human papilloma virus that causes common warts, hardly bother us.

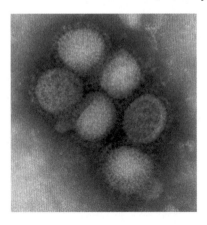

The H1N1 flu virus under magnification

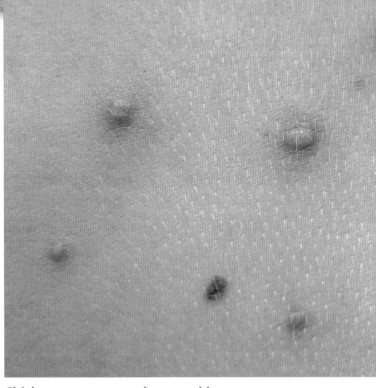
Chicken pox sores on human skin

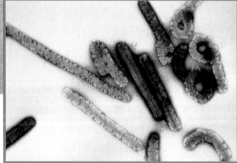
The Marburg virus

Antibiotics don't treat viral diseases; they only affect bacteria. That is why doctors don't prescribe antibiotics for the flu. At this time, vaccines (also known as immunizations) are the most powerful weapon we have against viral diseases. Vaccines are a form of prevention. They can be made of harmless bits of modified or dead virus pieces, which are usually injected into the muscle or blood. The body recognizes these virus pieces as "invaders" and uses the immune system to fight them, even though the virus in this form cannot make you sick. The immune system builds molecules called antibodies in response to the vaccine which can then respond quickly and kill off the real viral invaders before any damage occurs.

The vaccine for smallpox was the first vaccine developed and has been responsible for nearly wiping out this killer disease. Polio is another viral disease that has nearly been eliminated due to successful worldwide vaccination campaigns. In spite of vaccination, measles kills about 200,000 people each year and can cause miscarriages in pregnant women. But if we were to stop vaccinating for measles, it is estimated that about 2.7 million people would die each year.

A few antiviral drugs have been developed recently to treat HIV infections, influenza, and herpes. Scientists are working on additional antivirals, but progress is difficult because viruses live inside cells. That makes it hard to kill the virus without harming the host cell. Additionally, viruses' genetic material changes quickly, which means, for instance, that a vaccine for this year's flu might not be effective next year. That is why it is recommended to get a flu shot every year, to help give you protection against new forms of the virus.

Investigation 8: Diversity of Life

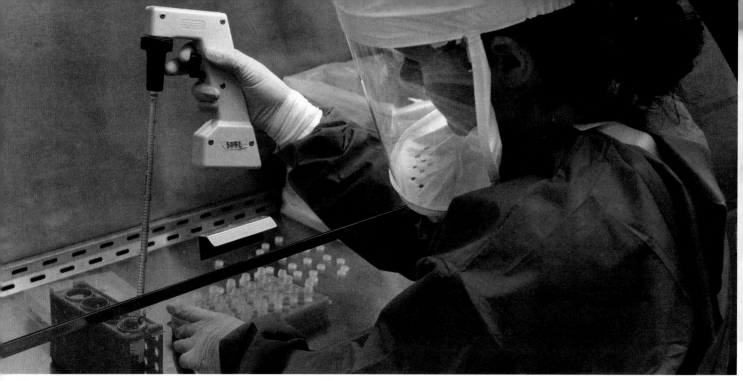

A scientist uses a pipette to transfer a virus into vials for sharing with other laboratories for health research.

Are Viruses Living?

Today, we are still just beginning to learn about viruses. Yes, viruses can kill humans. Viruses can also kill bacteria, including bacteria that are harmful to us. Viruses help maintain the ecological balance of organisms in the ocean, including those that produce oxygen for life on this planet, and are part of the human genome and every other living organism's genetic material.

And so we return to our question, one that scientists are actively discussing: "How do you know if something is living?" Do the characteristics of life, which we so carefully developed, apply to viruses? What is your conclusion?

Should viruses be placed in a new domain of life?

Think Questions

1. How do viruses depend upon cells?
2. How do you think virus genes became part of human genes?
3. How do humans protect themselves (and their animals) from viral diseases?
4. Are viruses living or nonliving? Support your conclusion with evidence from the article and from your studies in class.

Images and Data

Images and Data Table of Contents

Investigation 2: The Microscope
Microscope Parts **69**

Investigation 3: The Cell
Microorganism Guide **70**
How Big Are Cells? **74**

Investigation 4: Domains
Levels of Complexity Research Pages **76**
Archaea Family Album **79**
The Three Domains of Life **81**

Investigation 6: Plant Reproduction and Growth
Flower Information **82**
Flowers and Pollinators **86**

Investigation 7: Insects
Insect Structures and Functions **93**

References
Science Safety Rules **98**
Glossary **99**
Index **104**

Microscope Parts

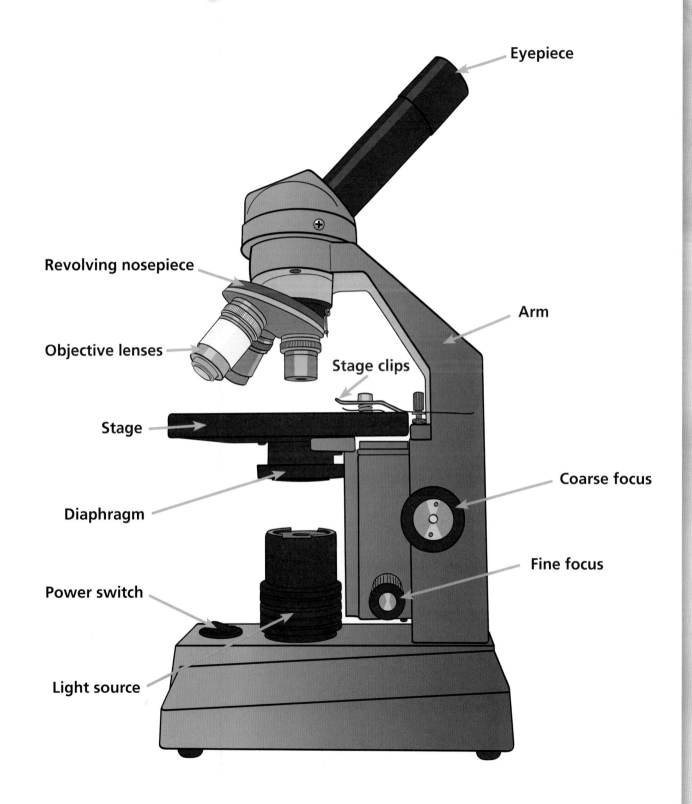

Investigation 2: The Microscope **69**

Microorganism Guide

Some of the organisms in this guide are considered pond organisms. Why might you find them in soil?

Green Algae

1. *Coelastrum*
Each cell is 7–10 µm wide

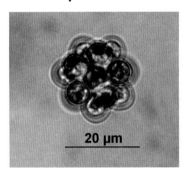

2. *Spirogyra*
50–100 µm

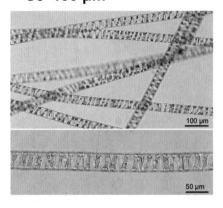

3. *Protococcus*
5–12 µm

4. *Cladophora*
300–1000 µm

5. *Hydrodictyon*
Up to 6 cm

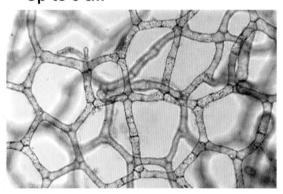

6. *Microspora*
Cell width 8–20 µm

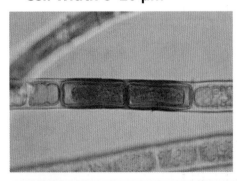

7. *Oedogonium*
Each cell about 20 µm long

Ciliates

1. **Didinium**
   50–100 µm

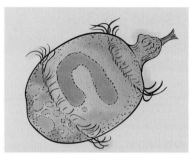

2. **Paramecium**
   50–330 µm

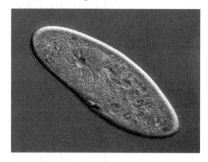

3. **Blepharisma**
   75–300 µm

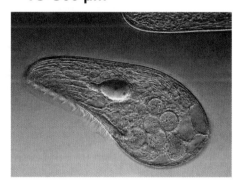

4. **Spirostomum**
   Up to 1 mm

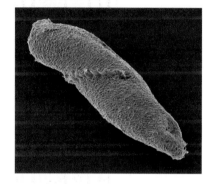

5. **Stentor**
   Up to 2 mm

6. **Euplotes**
   80–200 µm

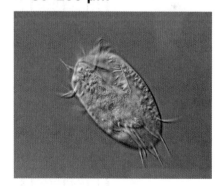

7. **Vorticella**
   100–200 µm long

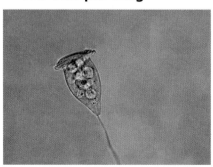

8. **Zoothamnium**
   Each cell is about 100 µm long

Investigation 3: The Cell

Flagellates and Sarcodines

Flagellates

1. **Chlamydomonas**
   10 µm in diameter

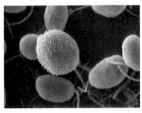

2. **Dinobryon**
   20 µm long

3. **Euglena**
   15–500 µm long

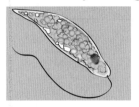

4. **Peridinium**
   10–100 µm long

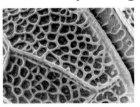

5. **Synura**
   Colony can be 50 µm across

6. **Volvox**
   Each cell – 4–8 µm, a colony can be 2 mm across

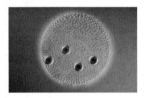

7. **Codosiga**
   5–10 µm long

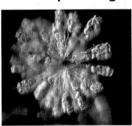

8. **Oikomonas**
   10–15 µm long

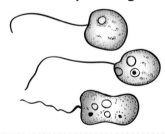

9. **Bodo**
   5–10 µm long

Sarcodines

1. **Amoeba**
   200–750 µm in diameter

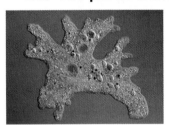

2. **Difflugia**
   Can be 200 µm

3. **Actinosphaerium**
   About 300 µm

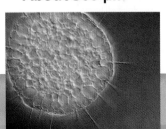

4. **Arcella**
   70–125 µm in diameter

Crustaceans, Rotifers, and Others

Crustaceans

1. *Daphnia*
1–5 mm

2. *Copepods*
.5 – >2 mm

3. *Fairy shrimp*
11–25 mm

Rotifers

1. *Rotaria*
100–500 µm

2. *Philodina*
100–500 µm

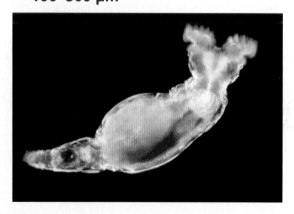

Others

1. *Nematode*
300 µm and up to 8 m

2. *Hydra*
5–20 mm long

3. *Tardigrade*
300–500 µm long

How Big Are Cells?

As a rule, cells are small. The illustration on the facing page shows the relative sizes of several cells that you might have seen. They are illustrated with a typical strand of human hair in the background for size comparison. As you can see, most cells are smaller than the diameter of a single hair.

The sizes of the cells are indicated in units called micrometers (μm). When you use a metric ruler in science class, the tiny marks on the ruler each indicate a millimeter. One millimeter is 1,000th of a meter. If you take that millimeter and divide it into 1,000 parts, you have divided it into micrometers. A micrometer is 1,000th of a millimeter, or 1,000,000th of a meter.

You can barely see objects that are 1 or 2 μm across if you use a good compound microscope set at high power (400X).

A human hair is about 100 μm in diameter. If you lay ten hairs side by side they will measure about 1 millimeter (mm). It might take 50 or more bacteria to equal the diameter of the hair because bacteria are generally 1–2 μm in size. However, some bacteria are significantly smaller—less than 1 μm.

The human cheek cell is a pretty large human cell at 40–50 μm in diameter. The little red blood cells are 5–7 μm, placing them on the small side of human cells. The majority of the 100 trillion cells in a human are in the 20 μm range, although the longest human cells (nerve cells in the spinal cord) can reach 1 meter (m) in length!

Human red blood cells

Some paramecia can grow as large as 300 μm, gaining them "elephant" status among their kin. However, they are tiny in comparison to the largest single-celled organisms discovered. In 2011, gigantic single-celled sponge-like protists called xenophyophores (zee·no·FL·oh·fors) were discovered in the Mariana Trench, 10,600 m below the surface of the Pacific Ocean. Many of these cells are more than 10 centimeters (cm) in diameter! You certainly don't need a microscope to see this cell.

The xenophyophore is a unique organism. Generally, cells are much smaller than 10 cm across and therefore are able to efficiently conduct the business of life. Small cells can easily circulate vital gases and food to all parts of the cell, and quickly move wastes to the cell membrane for removal. If cells were too large, cell structures in the center of the cell would not get the resources they need to continue functioning. This is the main factor limiting cell size.

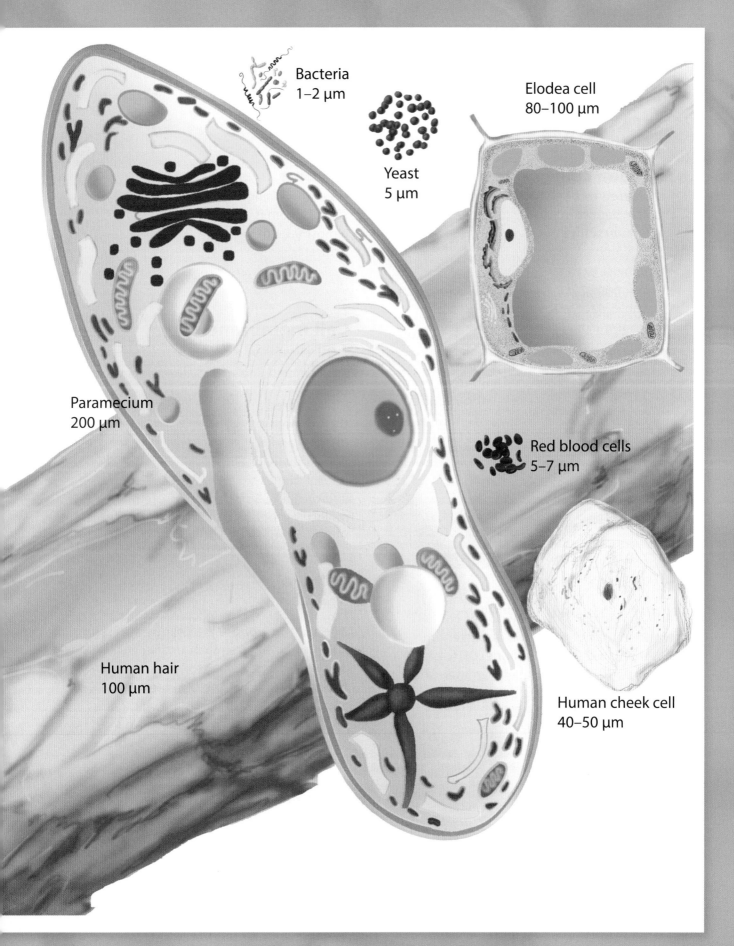

Levels of Complexity Research Pages

Archaea cell

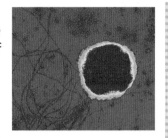

Considered to be the most ancient form of life known to live on Earth. Can be found in the most extreme environments thinkable. Archaeans vary in size from 0.1 to more than 15 micrometers (μm) long. They have genetic material called DNA, a cell wall, and a cell membrane that is different from any other organism's cell membrane. They do not have organelles such as a nucleus or mitochondrion.

Bacterium cell

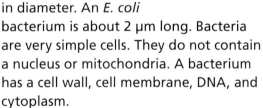

Found in three distinct shapes. A small spherical bacterium can be about 0.15–0.20 μm in diameter. An *E. coli* bacterium is about 2 μm long. Bacteria are very simple cells. They do not contain a nucleus or mitochondria. A bacterium has a cell wall, cell membrane, DNA, and cytoplasm.

Carbohydrate

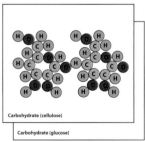

A family of molecules that includes sugars (such as glucose), starch, cellulose, and chitin. Carbohydrate molecules vary widely in size, from about 1 nanometer (nm) long to over 5,000 nm long. Carbohydrates are made of only carbon, hydrogen, and oxygen.

Carbon

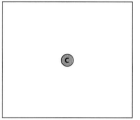

An atom. A single carbon atom is almost 0.2 nm in diameter.

Cell membrane

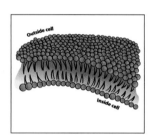

A cell structure found in most cells. It is considered the boundary of a cell and is made mostly of phospholipids and proteins. Some of the proteins allow molecules to travel into and out of the cell. A cell membrane is about 7 nm thick.

Cell wall

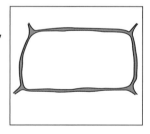

A cell structure found in bacteria, archaea, fungi, and plant cells. The cell wall is different in each of those organisms, but it always gives structure to the cell. It is built of complex carbohydrates such as cellulose, which makes up plant cell walls. It is about 2 nm thick and completely surrounds the cell membrane.

Chloroplast

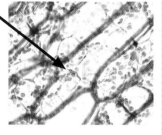

A large cell structure found in plants and algae (and some bacteria). Chloroplasts contain the green pigment chlorophyll, and as many as 100 can be found in some plant cells. It has a flat disk shape, its membrane is made of lipids and proteins, and it can range in size from 2–10 μm in diameter.

DNA

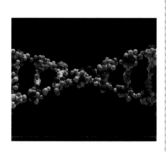

A complex molecule that is the genetic material of almost all life. It is a spiral-shaped chain composed of many units made of carbon, oxygen, hydrogen, nitrogen, and phosphorus atoms. Each unit is about 1 nm in size, but the entire molecule is twisted and packed so tightly that if stretched out, it could be a few centimeters long! Every species of organism has its own unique DNA code.

Elodea cell

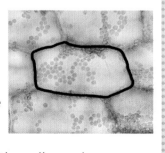

A plant found in freshwater ponds. Each cell is surrounded by a strong cell wall made of a carbohydrate called cellulose. Inside the cell wall is the cell membrane. Various cell structures are found in the cytoplasm, including a nucleus, vacuoles, chloroplasts, and mitochondria. Elodea cells can vary in size but, on average, are about 80–100 μm long.

Fungus cell

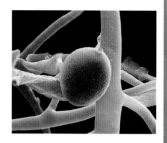

A form of life that includes yeast, mold, and mushrooms. There are many different kinds of fungi. A yeast cell can be as small as 2 μm in diameter. Fungus cells have cell walls made of chitin (a carbohydrate also found in insect exoskeletons). A fungus cell also has a cell membrane, nucleus, mitochondria, and other cell structures.

Human cheek cell

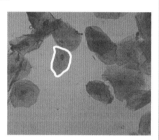

One of many kinds of cells found in humans. Each cell is about 40–50 μm in diameter. It is a typical animal cell and has numerous cell structures, including a cell membrane, a nucleus, and many mitochondria.

Hydrogen

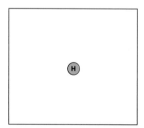

The smallest atom. It is about 0.1 nm in diameter.

Lipid

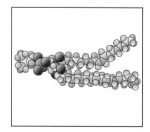

Also known as fat. One type of lipid is the phospholipid molecule, which forms the backbone of the cell membrane. It is less than 1 nm across. Phospholipids are made of carbon, hydrogen, and phosphorus.

Mitochondrion

A cell structure found in all cells except bacteria and archaea. Mitochondria are the "powerhouses" of cells that turn food into usable energy for the cell.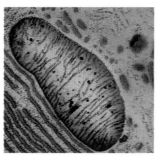
These organelles vary in size from 0.5–10 µm. Mitochondria have membranes made of phospholipids and proteins. They have their own DNA and in the distant past might have been independent organisms.

Nitrogen

An atom. It is about 0.15 nm in diameter.

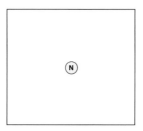

Nucleus

A cell structure found in all cells except bacteria and archaea. The nucleus contains the DNA of the cell. It is the largest organelle found in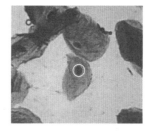
animal cells. The nucleus in a human cheek cell is about 10 µm in diameter. It has a membrane surrounding it that is made of phospholipids and proteins.

Oxygen

An atom. It is about 0.13 nm in diameter.

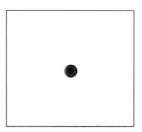

Paramecium

A free-living single-celled protist. Paramecia live in fresh water, but can survive in a dormant condition in soil.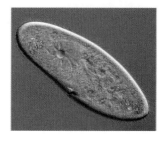
The *Paramecium caudatum* you saw in class is between 180 and 300 µm long. The paramecium has a cell membrane, nucleus, mitochondria, vacuoles, and other cell structures.

Phosphorus

An atom. It is about 0.25 nm in diameter.

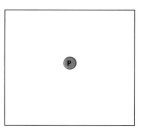

Protein

A molecule composed of smaller units called amino acids. Proteins vary in size but can be just several nanometers in diameter when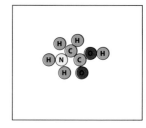
folded up. Proteins are important in cellular structures and in many chemical reactions within a cell. Proteins are made of carbon, hydrogen, oxygen, and nitrogen.

Archaea Family Album

Halobacterium salinarum

Halo means salt. A suitable environment for this halophile (salt lover) requires a high concentration of salt. The level of salinity needed for halobacteria to survive is far greater than that of the ocean, and would kill other nonhalophilic organisms, which means essentially every other organism on Earth.

H. salinarum is found in saline bodies of water like Mono Lake, the Great Salt Lake, and the Dead Sea. In fact, it is responsible for the bright red color that sometimes occurs in these bodies of salt water.

Scientists originally named these archaeans "halobacteria" before finding out they weren't bacteria at all!

Methanobrevibacter smithii

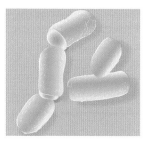

This archaeon is one of the microbes found in the human gut. That means it lives within the human digestive system and assists with digestion.

M. smithii breaks down complex sugars and produces methane gas. (Do you see a word similar to methane in its name?) In fact, *M. smithii* is responsible for the majority of methane production in humans!

Scientists are trying to learn more about how these organisms affect human digestion, because they have found that *M. smithii* may play a role in weight gain.

Methanocaldococcus jannaschii

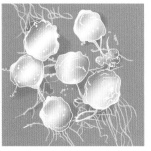

This archaeon is thermophilic (*thermo* = heat, *philic* = attracted to), and was discovered in a hot spring in Massachusetts. It produces methane gas. (Do you see a word similar to methane in its name?)

M. jannaschii was the first archaeon to have its complete DNA sequenced (analyzed). This was very important because scientists had classified archaea with bacteria, thinking they were all forms of bacteria. It wasn't until they could compare the DNA of organisms that they started to realize archaea and bacteria are two distinct major divisions (domains) of life.

Crenarchaeota

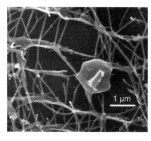

Crenarchaeota is a category of archaeans. Some crenarchaeotans live in extreme heat. They have been found growing in the highest temperature of any known organism—115°C. NASA scientists believe that hot springs may have once existed on Mars, and if there was life in them, it might have been similar to Crenarchaeota.

But crenarchaeotans have also been found in the bodies of organisms on the ocean floor and in the soil. Scientists suspect that crenarchaeotans in the soil may outnumber bacteria in the soil by a huge factor.

Sulfolobus solfataricus

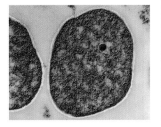

S. solfataricus is a thermophile and an acidophile, meaning it thrives in places that are extremely hot and acidic. This archaeon is found in almost all volcanic areas, including Yellowstone National Park and Mount St. Helens. It was named for the volcano Solfatara in Italy, where it was discovered.

These organisms metabolize sulfur found in the volcanic zones in which they thrive. (Do you see a word similar to sulfur in their name?) They live in volcanic hot springs or volcanic formations called mud pots, which are areas of boiling mud.

Methanococcoides burtonii

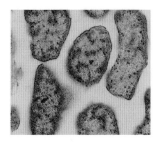

This archaeon is psychrophilic, (*psychro* = cold, *philic* = attracted to), meaning it thrives in extreme cold. *M. burtonii* lives at the bottom of Ace Lake in Antarctica, where the temperature hovers just 1 or 2 degrees above the freezing point of water.

The key to survival at these temperatures is the cell membrane. First, the proteins in their cell membranes are flexible instead of rigid. Second, they modify fats (phospholipids) in a way that makes them less likely to "freeze." These organisms produce methane gas. (Do you see a word similar to methane in their name?)

Pyrococcus furiosus

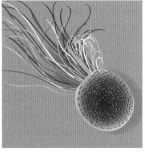

This archaeon is hyperthermophilic (*hyper* = excess, *thermo* = heat, *philic* = attracted to), which means it is in the small category of known organisms that thrive at temperatures above the boiling point of water, which would kill most living organisms. *P. furiosus* is found in hydrothermal vents in the ocean. The name *Pyrococcus* means "fireberry" in Greek, due to its round shape and extreme heat environment.

P. furiosus is also known for containing the element tungsten in its molecules, which is rare among organisms.

Cenarchaeum symbiosum

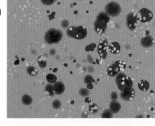

C. symbiosum lives in marine sponges which are organisms living in the ocean. These archaeans thrive at temperatures between 10°C and 60°C, which is a suitable environmental temperature for most organisms.

It is a symbiont of the sponge, meaning it cannot survive outside of the sponge.

The Three Domains of Life

Current classification of life is in three domains: Bacteria, Archaea, and Eukaryota. Bacteria and Archaea consist of prokaryotic organisms. Eukaryota consists of eukaryotic organisms.

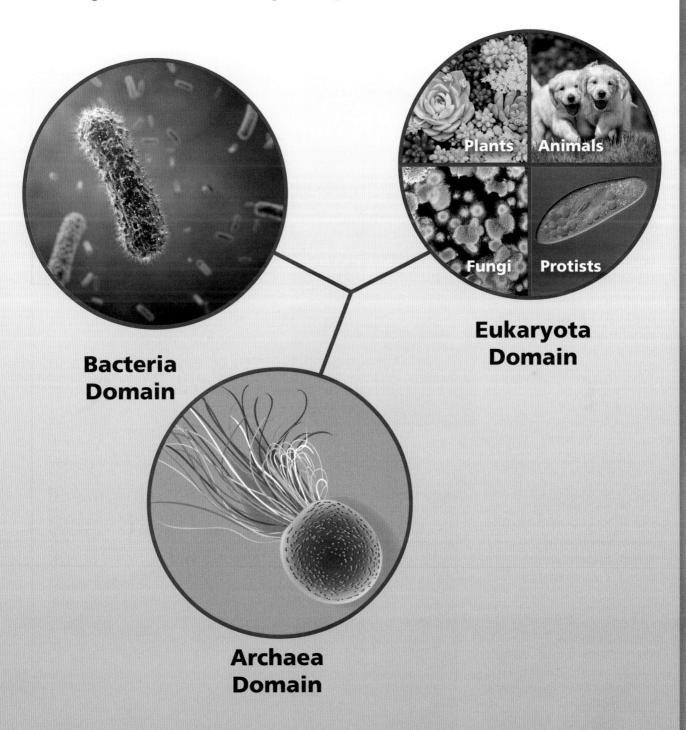

Flower Information

Camellia

The camellia is an evergreen shrub that has beautiful flowers that range from white to deep red. Camellia flowers are 5–10 cm across and stay open day and night.

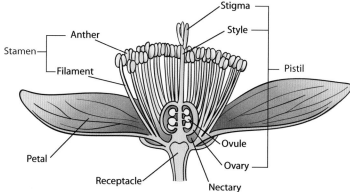

Fuschia

Fuschia flowers are usually red, pink, and lavender. They hang upside down from their stems like lanterns. Their large store of nectar is high up in a narrow tube, requiring the pollinator to reach 2–3 cm up into the flower.

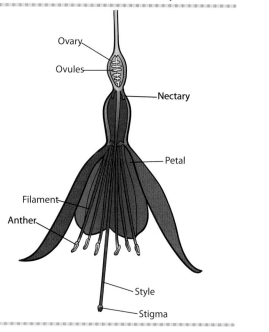

Snapdragon

The snapdragon produces many flowers on long stems. Each flower is about 3 cm deep and 1 cm across. The flower looks like a closed mouth with a large lower lip sticking out. To get into the flower for nectar, a pollinator must push the flower open.

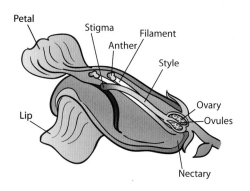

Tiger lily

The tiger lily's deep blooms are large enough for small pollinators to crawl into the flower to reach the nectar.

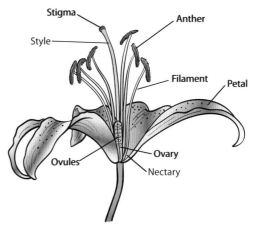

Firewheel

The firewheel is a composite flower with hundreds of tiny disk florets in the center. The open face of the flower makes it possible for pollinators with large or small wings to land on the flowers as they seek food. Some pollinators simply crawl from floret to floret, seeking nectar.

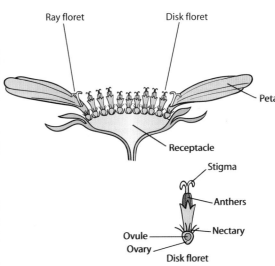

Daffodil

Daffodil flowers are usually bright yellow and fragrant. The pistil and stamens are inside a cup or trumpet-shaped structure that is about 2 cm wide. In order for a pollinator to get the nectar, it must crawl down into the cup.

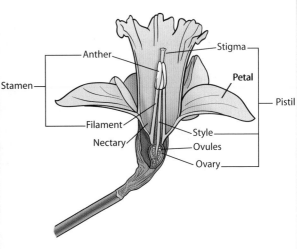

Investigation 6: Plant Reproduction and Growth

Poppy

Single poppies grow on the ends of stalks. The flowers are open and can receive flying pollinators and insects that climb up the stalks to the flowers. Poppies do not have a nectary.

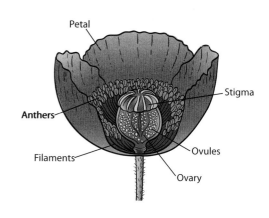

Wild rose

Wild roses are simple flowers, usually with five petals. They have several ovaries with the styles rising up in the center of the flower. Most roses we see at the store are bred for many more petals and colors. Roses do not have a nectary.

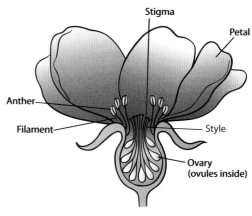

Penstemon

The penstemon flower is long and tubular. Many penstemons are red or pink, though common native flowers can be white, blue, or purple.

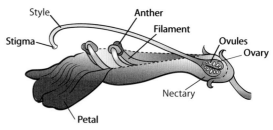

Tulip

Tulips have a sweet scent and brightly colored petals to attract pollinators.

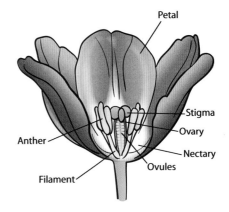

Alstroemeria

The alstroemeria is a native of South America. Its petals have stripes that act as guides to pollinators.

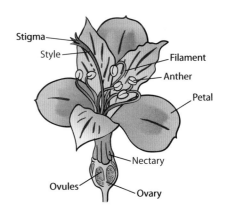

Flowers and Pollinators

Pollination syndromes

The diversity of flowers in the world is the result of the process of evolution. Fossil records show that plants that reproduce by seeds first developed almost 360 million years ago. Seeds allowed plants to survive in different environments on land. A seed can lie dormant until environmental conditions are optimum for survival. The first seed-producing plants depended on wind to get the pollen (sperm) to the egg. About 130 million years ago, flowering plants appeared. From this beginning, scientists estimate 250,000 to 400,000 species of flowering plants have developed.

Some flowering plants such as grasses still depend upon the wind for pollination. But most flowering plants have evolved relationships with animal pollinators, usually insects, to get the job done. Every species of flower that relies upon animal pollinators has a group of characteristics known as a pollination syndrome that attracts certain pollinators. Flowers and pollinators have changed together (**coevolved**) over millions of years to enable the success of each kind of organism. However, if natural variation (differences) among plants leads to flowers that are a little different color or shape, or if they bloom earlier in the season than the previous generation, the plant may no longer attract pollinators. Those flowers might not reproduce and that species of flower could die out. If the pollinators evolve with the flowering plant's changes, both kinds of organisms will continue to succeed.

What determines which type of flower a pollinator chooses?

Flowering plants employ one or more of several strategies to attract pollinators: visual cues (such as shape and size), color, scent, food, timing, mimicry (imitation), and entrapment.

Shape and size of flower

Flower shape is important to pollinators. Many flowers contain nectar in a structure called a nectary, which may be deep within the flower. If a pollinator is looking for nectar, it must have a way to reach it. Some pollinators, such as bees, have the ability to travel from flower to flower, while others, such as beetles, are less agile and prefer flowers that each deliver a lot of nectar and pollen.

Pollinator	Flower shapes
Ants	Small, low-growing flowers that are close to the stem
Bats	Large and bell-shaped
Bees	Flowers that only open when the relatively heavy bee lands on its petals, so lighter pollinators can't gain access Symmetrical flowers where one side is a mirror image of the other side Nectary at the base of the tube or cup
Beetles	Bowl-shaped flowers, open flowers that provide a stable landing site and space for beetles to crawl around and chew on the flower Large solitary flowers or large clusters of small flowers
Butterflies	Flat-topped flowers or cluster of flowers Landing platforms where butterflies can perch and probe into the flower for nectar using their long proboscis (tongue)
Flies	Funnel-shaped
Hummingbirds	Trumpet, bell-shaped, funnel, and cup-shaped flowers (Hummingbirds hover above the flower and use their long beaks to extract nectar.)
Moths	Open, without a lip

Investigation 6: Plant Reproduction and Growth

Color of flower

Color signals the pollinator that there is a reward for visiting the bloom. Most often that reward is nectar or pollen. Flowers that rely on color may display a vibrant color that pinpoints the location of the nectar. Other flowers reflect ultraviolet light, which some insects can see. Some flowers change color after a pollinator has visited them, making them virtually invisible to other pollinators. They no longer need to attract a pollinator.

Pollinator	Attractive colors
Bats	White and pastels
Bees	Yellow, blue, purple, ultraviolet Color patterns direct the bee to the ideal landing spot
Beetles	White, green
Butterflies	Red, orange, yellow, pink, purple, bright colors
Flies	Green, lime, white, cream, dark brown, purple, maroon
Hummingbirds	Red, orange, purple-red, yellow
Moths	White, green, dull colors

Scent

Scent can powerfully attract pollinators that either cannot see color or are active at night when colors are difficult to see. Flowers that employ scent as an attractant are usually drab in color like dark-red, purple, or brown, or they are pale or white. The scents in some flowers are so strong they can be detected from as far away as 1 kilometer (km).

Pollinator	Attractive scents
Bats	Very fragrant tropical fruit and fermenting fruit aromas, herbs
Bees	Sweet or minty fragrance
Beetles	Strong, fruity scents Fetid (extremely unpleasant)
Butterflies	Faint but fresh
Flies	Rotting meat, dung, earthy smells, and blood
Hummingbirds	None (Birds have no sense of smell and are not attracted by scents at all.)
Moths	Female moth pheromone scents, strong flower scents Flowers that emit more scent at night than during daylight

Food

The main reason animals visit flowers is for food. Pollen and nectar are the main food for pollinators such as bees, beetles, butterflies, and hummingbirds. The draw of a meal is sometimes all that a plant needs to offer. The small, green flowers of grapevines may not seem to be attractive, but an abundance of nectar exposed at the surface of the flower attracts short-tongued bees, flies, and wasps. Plants that conceal their treasure deep within the flower typically use scent and color to advertise the presence of nectar. Plants that attract pollen eaters typically produce enough pollen to satisfy the pollinator as well as assure that some of the pollen is carried away to pollinate another flower.

Pollinator	Attractive foods
Bats	Lots of nectar and pollen
Bees	Nectar and pollen
Beetles	Nectar
Butterflies	Nectar hidden within the flower
Flies	Pollen and exposed nectar
Hummingbirds	Nectar deep within the flower
Moths	Lots of nectar hidden within the flower
Wasps	Exposed nectar

Timing (day/night and seasonal)

Pollinators that are active at night, such as bats and moths, need to find flowers that are open at night. Night-blooming plants can't rely on typical visual cues, so they tend to have large, pale flowers that are very fragrant to attract pollinators. Some flowers may only be open for a short time during the day, such as morning glories that open early in the morning when fewer flowers are open on other plants.

Timing can also be seasonal. Different plants flower at different times throughout spring, summer, and fall. A flowering plant can be assured of a steady stream of pollinators throughout its blooming season. Flowers that open early in spring are often smaller because after a long winter, pollinators are eager for a meal of pollen and nectar. Summer flowers tend to be large, brightly colored, and fragrant.

Pollinator	Attractive times
Bats	Night, closed during the day
Bees	Day
Beetles	Day
Butterflies	Day
Hummingbirds	Day
Moths	Late afternoon or night

Investigation 6: Plant Reproduction and Growth

Mimicry (imitation)

Many species produce a flower and/or scent that attracts the pollinator but does not return the promised rewards to the pollinator. Some flowers have a scent that mimics or imitates rotting meat. An insect attracted to such a smell will lay its eggs, and in the process pollinate the flower. When the eggs hatch, however, there is no rotting flesh for the larvae to eat, and they die.

Orchids are the masters of mimicry. One type of orchid appears to be a female wasp and emits chemicals that attract male wasps. When the male wasp attempts to mate with the "wasp," he picks up pollen that he transfers to the next flower.

Note that most flowers you observed in class do not use mimicry to attract pollinators.

Pollinator	Attractive mimicries
Bees	Bee-shaped flowers
Flies	Food (scent of insect prey, rotting meat)
Moths	Female moth pheromone scent
Wasps	Wasp-shaped flowers and pheromones

Entrapment

A few plants have evolved a different way to accomplish pollination. The plant attracts the pollinator by a combination of cues, including appearance, scent, food, and mimicry. As the pollinator seeks its reward, it is trapped within the flower. In efforts to escape the flower, it becomes covered in pollen, which it will carry away when it eventually gets free.

Insect Structures and Functions

All insects are multicellular animals without a backbone (invertebrates). They all have

- an exoskeleton (a hard surface covering an insect);
- segmented bodies with three major sections: head, thorax, abdomen;
- three pairs of jointed legs (six legs total);
- one pair of antennae; and
- simple and compound eyes.

The exoskeleton provides protection for internal organs, anchors the muscles, and keeps the insect from drying out. An insect's exoskeleton is made of a strong, lightweight substance called chitin (KY•tin), which is also the base material in other animals' horns and is similar to the material that makes up human fingernails.

While all insects share this structural design, their body parts vary greatly, depending on different survival needs. Variation in these fundamental structures has produced the most diverse collection of animals on Earth. In fact, it is estimated that there are more than 1 million species of insects!

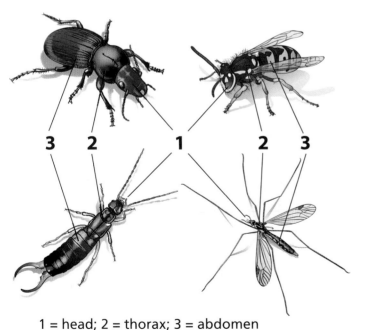

1 = head; 2 = thorax; 3 = abdomen

Structure/Behavior/Function Definitions

structure a tissue, organ, or other formation made up of different but related parts

behavior a manner of acting

function the specific activity performed by an organ or part; the purpose of a behavior

Head: Eyes, Antennae, and Mouthparts

Take a moment to examine the head of your cockroach. You will notice three distinct structures: eyes, antennae, and mouthparts.

Eyes. Eyes provide insects with information about their environment. Insects have two kinds of eyes—simple and compound. Two large compound eyes are made of many small lenses (up to 25,000) that detect color and motion. Scientists compare looking through compound eyes to watching a thousand TV screens at once, with each screen showing an image of the object from a slightly different angle. Simple eyes are smaller and often found on the forehead. They register changes in light intensity, allowing insects to detect day length. This helps insects determine seasons and thus program their bodies to prepare for reproduction, migration, hibernation, and other activities.

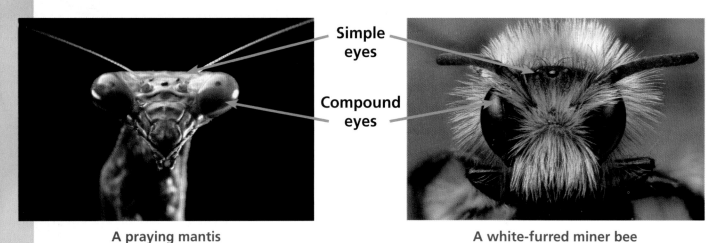

A praying mantis — Simple eyes, Compound eyes

A white-furred miner bee

Antennae. Insects always have two antennae, usually positioned near the eyes. These movable parts allow insects to sense odors, vibrations, and other information about their environment. Antennae come in a huge range of shapes and sizes, and may even differ between genders. There are five main types of antenna structures.

Type	Bristle-like	Clubbed	Elbowed	Feather-like	Thread-like
Examples	Dragonflies	Butterflies, moths, beetles	Ants, beetles, bees	Moths, mosquitoes	Ground beetles

Mouthparts. Insect mouthparts can tell us a lot about their feeding habits. The shape of an insect's mouthparts is adapted to a specific kind of feeding, helping the insect get the nutrients it needs.

	Chewing	Piercing/sucking	Sponging	Siphoning
Type	Chewing	Piercing/sucking	Sponging	Siphoning
Examples	Beetles, grasshoppers	Mosquitoes, aphids, true bugs	Housefly, blowflies	Moths, bees, butterflies

The beetle's mouthparts allow it to grasp and chew food.

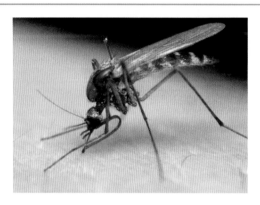

The mouthparts of a mosquito allow it to pierce skin and suck blood out of its host.

What structure do you notice here that would help a housefly sponge up liquids?

The moth's siphoning mouthpart helps it suck nectar from the flower.

Investigation 7: Insects

Thorax: Wings and Legs

The thorax of all insects is divided into three distinct segments. One pair of legs is attached to each segment of the thorax. Can you see these segments on your cockroach?

Wings. Insects are the only group of invertebrates known to have evolved the ability to fly. Most insects have two pairs of wings attached to the thorax. In some groups of insects (such as beetles), the front pair of wings has evolved into a hard covering, protecting the second pair of wings, thorax, and abdomen. Other insects have ridges on their wings that produce sound when rubbed together, producing the familiar chirping noise of the cricket and the drone of the cicada.

Type	Hardened outer wing	Flying wings and halteres (modified wings)	Membrane-like	Scales
Examples	Beetles	Mosquitoes, flies	Dragonflies, bees, wasps, termites	Moths, butterflies

A scarab beetle

A housefly

A dragonfly

An eastern tailed blue butterfly

Legs. The legs of insects are greatly varied, as they have adapted for different kinds of movement. Many insects have uniquely shaped hooks, spines, and bristles on their legs for holding onto twigs and leaves. These also are useful for frequent grooming of the insect's eyes, face, and antennae to ensure that their sensory tools remain in prime condition. Flies have sticky pads on their feet that allow them to walk up smooth surfaces like glass.

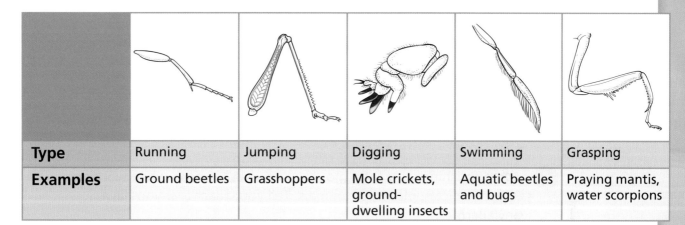

Type	Running	Jumping	Digging	Swimming	Grasping
Examples	Ground beetles	Grasshoppers	Mole crickets, ground-dwelling insects	Aquatic beetles and bugs	Praying mantis, water scorpions

Abdomen

The abdomen contains the guts of the insect. Here you will find a modified heart, intestines, and reproductive organs. In many animals, the circulatory system carries oxygen and nutrients to each cell. In insects, the transport of oxygen and nutrients is separate.

Insect blood doesn't carry oxygen. It flows around the gut, where it picks up nutrients from the digested food and carries the nutrients to the cells. It also takes away waste products.

Insect cells get oxygen from a network of tracheae. These hollow tubes branch out to provide oxygen to every cell in an insect's body. The tracheae are connected to outside air by openings on the abdomen called spiracles.

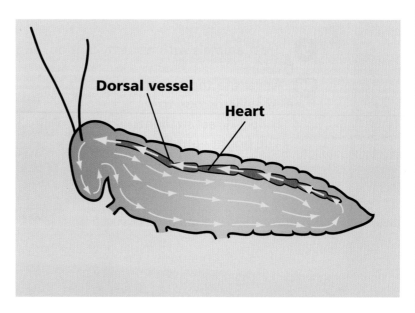

Open circulatory system for delivering nutrients

Science Safety Rules

1. Always follow the safety procedures outlined by your teacher. Follow directions, and ask questions if you're unsure of what to do.

2. Never put any material in your mouth. Do not taste any material or chemical unless your teacher specifically tells you to do so.

3. Do not smell any unknown material. If your teacher asks you to smell a material, wave a hand over it to bring the scent toward your nose.

4. Avoid touching your face, mouth, ears, eyes, or nose while working with chemicals, plants, or animals. Tell your teacher if you have any allergies.

5. Always wash your hands with soap and warm water immediately after using chemicals (including common chemicals, such as salt and dyes) and handling natural materials or organisms.

6. Do not mix unknown chemicals just to see what might happen.

7. Always wear safety goggles when working with liquids, chemicals, and sharp or pointed tools. Tell your teacher if you wear contact lenses.

8. Clean up spills immediately. Report all spills, accidents, and injuries to your teacher.

9. Treat animals with respect, caution, and consideration.

10. Never use the mirror of a microscope to reflect direct sunlight. The bright light can cause permanent eye damage.

Glossary

abdomen the third section of the insect body, including the digestive and reproductive organs and most of the circulatory and respiratory systems

adaptation any structure or behavior of an organism that allows it to survive in its environment

aerobic cellular respiration a process by which organisms convert glucose into usable energy

alga (plural **algae**) an aquatic protist containing chlorophyll. Algae may be single-celled or multicellular.

antibiotic a medicine that can kill many types of bacteria

aquatic living or occurring in water

archaea a microscopic, single-celled organism that lacks a nucleus and organelles (prokaryotic). Archaea have different cell walls and cell membranes than bacteria or eukaryotes.

asexual reproduction the production of genetically identical offspring from a single parent

atom a particle that is the basic building block of matter

bacterium (plural **bacteria**) a microscopic, single-celled organism that lacks a nucleus and organelles (prokaryotic).

behavior a manner of acting

biodiversity the variety of life that exists in a particular habitat or ecosystem

cell the basic unit of life. All organisms are cells or are made of cells.

cell membrane the boundary between a cell and its environment

cell structure a part of a cell with a specific job that enables an organism to carry out life's functions

cell wall a semirigid structure that surrounds cells of plants, fungi, and bacteria

chlorophyll a green pigment in chloroplasts that captures light energy to make sugars during photosynthesis

chloroplast an organelle containing chlorophyll, found in plant cells and some protists

cilium (plural **cilia**) (SILL•ee•uh) a short hairlike structure that propels protists through their fluid environment

classification a system or way of organizing living things

coevolve when two or more species affect each other's evolution

colony a group of organisms of the same species living together. A bacterial colony is a visible group of bacteria.

compound microscope a microscope that uses two lenses (eyepiece and objective lens)

contractile vacuole an organelle found mostly in protists that collects extra water in a cell and expels it

control an experimental test used to compare results with tests where a variable was changed

cotyledon the white, starchy part of a flowering plant seed. The cotyledon contains food to nourish the embryo during germination.

culture a growth of organisms on a prepared material

cuticle a waxy covering that covers leaves, reducing water loss through evaporation

cytoplasm all of the interior of a cell outside the nucleus

daughter cell a cell created during cell division that is an exact copy of the original

dead no longer alive

decomposer an organism that breaks down dead material and returns nutrients to the soil

digestive enzyme a chemical that breaks down food

dispersal the process of spreading out from a starting place

domain one group in the most currently accepted biological classification system. The three domains are Bacteria, Archaea, and Eukaryota.

dormant a state of suspended activity. Dormant organisms are alive but inactive.

ecosystem a system of organisms and environmental factors

egg the female sex cell

elodea an aquatic plant that grows in freshwater ponds and slow-moving streams

embryo the early developmental stage of a plant or animal

endoplasmic reticulum a cell structure involved in making proteins

energy the capacity to do work. Most energy used by organisms comes from the Sun.

environment the area in which an organism lives

environmental factor a condition of the environment that affects how suitable it is for a living thing

eukaryote an organism made of a cell or cells that contain a nucleus and organelles. All cells except bacteria and archaea are eukaryotic.

evidence information gathered by observation or experimentation

evolution changes to a species' genes over time (many generations) as different genes are passed from parent to offspring

evolve how a species changes over time (many generations) as different genes are passed from parent to offspring

fertilization the union of the nucleus of an egg cell with the nucleus of a sperm cell to produce a cell that will divide to become a new organism of the same type as the parent cells

field of view (FOV) the diameter of the circle of light seen through a microscope

flower the part of a seed plant that includes the reproductive organs

food a substance that provides energy and nutrients for organisms. Organisms use food for growth, repair, and cellular processes.

food-borne illness any illness resulting from the consumption of contaminated or poisonous food

fruit the ripened ovary of a plant, containing the seeds

function the specific activity performed by an organ or part; the purpose of a behavior

fungus (plural **fungi**) a eukaryotic organism, including molds, mushrooms, and yeasts. Can be single-celled or multicellular.

gas exchange one of the characteristics of life. Gas exchange occurs at the cellular level, with carbon dioxide, oxygen, and water vapor being the most common gases exchanged.

genetic factor genes in an organism's DNA

genetic material codes for the characteristics of organisms. Passed on from one generation to the next. Found in the form of deoxyribonucleic acid (DNA) or ribonucleic acid (RNA).

germinate the start of growth and development of a seed

growth increase in size of an organism. Growth is a characteristic of life.

guard cell a specialized plant cell that controls the opening and closing of the stomata, thus regulating transpiration

habitat a specific place where an organism lives

insect a class of animals with three body parts (head, thorax, and abdomen), six legs, and antennae

larva the immature, wingless, feeding stage in the life cycle of many insects

living the condition of being alive

lysosome an organelle in animal and protist cells that digests cellular waste

magnification the amount by which an object is magnified by a lens

magnify to make something appear larger than it actually is

microorganism an organism so small that a microscope must be used to view it

microscope an instrument used for viewing very small objects

mitochondrion an organelle that uses aerobic cellular respiration to change glucose into usable energy for the cell. Found only in eukaryotes.

molecule a particle made of two or more smaller particles held together by chemical bonds

multicellular organism an organism made of more than one cell

nonliving referring to something that has never been alive

nucleus an organelle that regulates protein production and contains genetic material

oral groove a fold leading to the food vacuole in some single-celled organisms

organ a structural unit made up of tissues that serves one function in a multicellular organism

organ system a group of organs that works together for one purpose in a multicellular organism

organelle a membrane-bound structure inside eukaryotic cells that performs specialized functions

organism an individual living thing, such as a plant, animal, fungus, bacterium, archaeon, or protist

ovary the part of the plant at the base of the pistil that contains the egg. After fertilization, the ovary turns into a fruit.

ovule a potential seed found within the ovaries of a plant

paramecium (plural **paramecia**) (pair•uh•ME•see•uh) a ciliated protist that lives in fresh water and eats other tiny organisms for food

pheromone a chemical released by an animal to communicate with or influence another organism

phloem (FLO•em) tissue within a vascular plant that transports food made in the leaves to all other parts of the plant

photosynthesis the process by which organisms that have chloroplasts use light energy, carbon dioxide, and water to make sugar

pistil a female reproductive structure in a flower. It consists of the ovary, containing the seeds, and the stigma.

plasmid circular pieces of genetic material (DNA)

pollen the tiny particles that contain the male sex cells. Pollen develops on the anthers.

pollen tube a tube through which a sperm travels to fertilize an egg in a flowering plant

pollination the transfer of pollen from the anther (male part) of a plant to a stigma (female part) of a plant, allowing fertilization of an egg

pollination syndrome a group of characteristics of a flower that has evolved to attract pollinators to help the plant successfully reproduce

pollinator an organism that transfers pollen from the anther (male part) of a plant to a stigma (female part) of a plant

power the amount a lens magnifies an object viewed through a microscope

prokaryote a single-celled organism that has no nucleus or organelles. All prokaryotes are bacteria or archaea.

protist (PRO•tist) eukaryotic, usually a single-celled organism

reproduce to create new individual organisms of the same kind. Some reproduce asexually (without the joining of two cells), and others reproduce sexually (the joining of egg and sperm cells).

response an organism's reaction to its environment

ribosome a cell structure involved in making proteins in all cells

root the underground part of a plant that functions as an organ to take up water and minerals, store food, and anchor the plant

root hair an extension of a cell near the root tip that takes in water and minerals

salinity the amount of salt in a substance

salt tolerant a characteristic of some plants that allows them to germinate and grow in salty environments

scale the proportional size of a magnified image compared to the original

seed a young plant in a dormant or resting stage, capable of growing into an adult plant

seed-dispersal mechanism a structure or feature of a seed that allows it to travel some distance from a parent plant

seed-dispersal strategy a way that seeds can travel away from the parent plant, such as wind or animals

sexual reproduction the creation of offspring when genetic material from two parents (in the form of an egg and a sperm) is combined.

single-celled organism an organism made of one cell that carries out all the functions of living. Also known as a unicellular organism.

species a unit of biological classification that refers to one kind of organism

sperm the male sex cell

spiracle an opening on the side of an insect that allows for gas exchange (oxygen enters and carbon dioxide exits).

spore a reproductive cell distributed through the air

stigma the tip of the pistil, which is often sticky and receives the pollen grain

stomata (singular **stoma**) openings on the surface of leaves that allow gas exchange. Guard cells control the opening and closing of the stomata.

structure a tissue, organ, or other formation made up of different but related parts

sugar one type of chemical compound produced by plants as a result of photosynthesis. Sugars are sources of energy for living organisms.

tissue material in a multicellular organism composed of similar cells that work together for a purpose

transpiration the process by which water flows through plants, entering the roots and exiting the stomata

vacuole a fluid-filled membrane in the cytoplasm of plant cells, fungus cells, and protist cells.

vascular system a group of tubes that carry sugars and water to all parts of a plant

vein a tube within an organism that is part of the vascular system of the organism

virus a microscopic agent that can invade cells of organisms and replicate. Scientific debate continues as to whether viruses are living or nonliving.

waste solids, liquids, or gases that are unusable by the cells of organisms and must be moved out of the cell

xylem (ZY•lem) a tissue made of long connected cells within a vascular plant that transports water and minerals from the roots to all the cells in the plant

yeast a single-celled fungus

Index

A
abdomen, 56, 99
adaptation, 5, 99
aerobic cellular respiration, 38, 39, 99
alga, 37, 99
animal, role in seed dispersal, 46–47
antibiotic, 22, 28, 99
aquatic, 18, 99
archaea, 15, 17, 63, 99
asexual reproduction, 22, 54, 99
atom, 9, 99

B
bacterium, 15, 17, 20–25, 26–30, 63, 65, 99
behavior, 13, 99
Belly Button Biodiversity Project, 24
Binnig, Gerd, 9
bioblitz, 59–60, 62
biodiversity, 59–62, 99

C
cell, 10, 14–19, 32, 38, 65, 99
cell membrane, 11–12, 16, 17, 18, 19, 21, 99
cell structure, 11, 99
cell wall, 16, 18, 19, 21, 22, 99
chlorophyll, 37, 99
chloroplast, 18, 99
cilium, 11, 99
classification, 79, 81, 99
coevolve, 99
colony, 21, 99
compound microscope, 8, 99
contractile vacuole, 12, 17, 99
control, 41, 99
cotyledon, 50, 100
culture, 21, 100
cuticle, 32, 100
cytoplasm, 16, 17, 18, 19, 21, 35, 100

D
daughter cell, 13, 100
dead, 4, 100
decomposer, 29, 100
digestive enzyme, 12, 100
dispersal, 43, 100
domain, 66, 100
dormant, 6, 100

E
ecosystem, 24, 100
egg, 49, 100
elodea, 37, 100
embryo, 49, 100
endoplasmic reticulum, 17, 18, 19, 100
energy, 5, 6, 12, 35–39, 100
environment, 5–7, 13, 100
environmental factor, 40, 100
Escherichia coli, 21, 24, 26–27, 29, 30
eukaryote, 66, 100
evidence, 20
evolution, 30, 100
evolve, 34, 51, 63, 100

F
fertilization, 50, 100
field of view, 15, 100
flower, 49, 100
food, 4, 12, 18, 39, 40, 100
food-borne illness, 27, 101
fruit, 5, 101
function, 4, 17, 101
fungus, 61, 63, 101

G
Galilei, Galileo, 8
gas exchange, 4, 6, 18, 37, 101
genetic factor, 17, 41, 101
genetic material, 16, 17, 101
germinate, 41, 101
growth, 6–7, 101
guard cell, 31, 36, 39, 101

H
habitat, 5, 101
Hooke, Robert, 9, 15
Human Microbiome Project, 25

I
insect, 14, 51–58, 101

J
Janssen, Zacharias, 8

K
Knoll, Max, 9

L
larva, 52, 101
leaf, 32–34, 41
Leeuwenhoek, Antoni van, 8, 9, 10, 16, 21
living, 3–7, 63–66, 101
lysosome, 17, 18, 19, 101

M
Madagascar hissing cockroach, 51–52
magnification, 8, 101
magnify, 8, 101
microorganism, 18, 101
microscope, 8–9, 11, 15, 21, 101
mitochondrion, 11, 17, 18, 19, 101
molecule, 21, 101
multicellular organism, 14, 101

N
nonliving, 4, 63–66, 101
nucleus, 17, 18, 19, 101

O
oral groove, 12, 101
organ, 11, 101
organ system, 17, 101
organelle, 11, 101
organism, 3–7, 10, 14–16, 101
ovary, 49, 102
ovule, 49, 50, 102

P
paramecium, 5, 10–13, 15, 102
pheromone, 56–58, 102
phloem, 38, 39, 102
photosynthesis, 31, 32–33, 34, 37, 38, 39, 102
pistil, 49, 102
plasmid, 16, 21, 22, 102
pollen, 49, 102
pollen tube, 49, 102
pollination, 49, 102
pollination syndrome, 102
pollinator, 61, 102
power, 8, 102
prokaryote, 17, 21, 102
protist, 10, 15, 102

R
reproduce, 6–7, 13, 49–58, 102
response, 6–7, 13, 102
ribosome, 16, 17, 18, 19, 21, 102
Rohrer, Heinrich, 9
root, 34, 38, 102
root hair, 36, 102
Ruska, Ernst, 9

S
salinity, 40, 102
Salmonella, 27
salt tolerant, 40–42, 102
scale, 15, 102
Schleiden, Matthias, 16
Schwann, Theodor, 16
seed, 40, 43–48, 103
seed-dispersal mechanism, 43, 46–47, 48, 103
seed-dispersal strategy, 43, 44–46, 48, 103
sexual reproduction, 13, 49–50, 103
single-celled organism, 10, 103
species, 6, 103
sperm, 49, 103
spiracle, 52, 103
spore, 6, 103
Staphylococcus, 27, 29
stigma, 49, 103
stomata, 31–32, 36, 39, 103
structure, 9, 11, 17, 18, 38, 43, 51, 64, 103
sugar, 37, 38, 39, 103

T
tissue, 27, 103
transpiration, 32, 36, 103

V
vacuole, 11, 18, 41, 103
vascular system, 36, 39, 103
vein, 36, 103
virus, 63–66, 103

W
waste, 4, 6, 18, 103
Wilson, E. O., 60, 62

X
xylem, 36, 38, 39, 41, 103

Y
yeast, 12, 103

Z
Zernike, Frits, 9